J. B. Killebrew

Oil Region of Tennessee

With some Account of its other Resources and Capabilities

J. B. Killebrew

Oil Region of Tennessee
With some Account of its other Resources and Capabilities

ISBN/EAN: 9783744758864

Printed in Europe, USA, Canada, Australia, Japan

Cover: Foto ©berggeist007 / pixelio.de

More available books at **www.hansebooks.com**

OIL REGION

OF

TENNESSEE

WITH SOME ACCOUNT OF ITS OTHER

RESOURCES AND CAPABILITIES.

By J. B. KILLEBREW,

Commissioner of Agriculture, Statistics and Mines.

NASHVILLE, TENNESSEE:
PRINTED BY "THE AMERICAN" PRINTING COMPANY.
1877.

The work which is herewith submitted, has been ready for the press for two months or more, but its publication has been delayed in order to prepare a suitable map to accompany it.

At the time of my investigations—in April and May—the demand for information pertaining to the oil districts of Tennessee came almost daily. Since that time the oil production in Pennsylvania has been greatly increased. In March the production per day was 29,087 barrels. In April this was increased to 32,427 barrels; May, 36,374; June, 37,693, the largest daily production ever obtained. The entire production for June was 1,130,790 barrels. This has had the effect of lowering the price of oil, and to prevent new developments. Nevertheless, the oil region of Tennessee, aside from its probable importance in the future as an oil producing district, has many attractions for the immigrant. Unlike almost every other locality where oil is found, it is a fruitful region, abounding in generous soils, excellent timber, unsurpassed water powers, and a considerable amount of mineral wealth. It is a high, healthy country, with pure air, sheltered nooks, sunny slopes, rug-

4

ged steeps. 1t has its plateaus, grand forests, sweet herbage, cooling springs, a scenery varied and picturesque, and is coursed by many beautiful streams, which give animation to the landscape, and supply, during a portion of the year, the means of transporting the products of the soil to market. I have, therefore, not confined myself simply to the oil indications but have given such general facts as may interest those seeking to found colonies, or to occupy a virgin soil. The continuation of the Tennessee and Pacific Railroad, or the McMinnville and Manchester Railroad to the Cincinnati Southern, would make this hitherto almost inaccessible region one of the fairest and most productive portions of the State.

I have the honor to be, Governor,

Your obedient servant,

J. B. KILLEBREW.

August 4, 1877.

TENNESSEE OIL REGION.

CHAPTER I.

PETROLEUM.

HISTORY, CONDITIONS OF DEVELOPMENT AND STORAGE, PRODUCTS AND EDUCTS, ORIGIN, USES—STATISTICS OF PRODUCTION.

Petroleum is no new product. It has been known, and used to some extent, in the arts for four thousand years. It was employed in the mortar at the building of Babylon and Ninevah. The Egyptians used it in embalming their dead. From time immemorial the bituminous matter of the Dead Sea has been known. The petroleum springs that ooze out upon the banks of the Is, a tributary of the Euphrates, attracted the attention of Alexander, and of several of the Roman Emperors. On one of the Ionian islands there is a spring which has been known for more than 2,000 years. In Zante, in Ecbatana, in Sicily, in Italy, on the Caspian Sea, in Persia, on the banks of the Irawaddy, in Bavaria, France, England, Scotland, in the Indian Archipelago, and in the Island of Trinidad, petroleum is found. It has been imported into Liverpool from Africa, and even China, with its universal resources and inventions, lays claims to unlimited supplies of oleaginous wealth.

But the largest quantities yet met with in the world are in the Northern States of America. New York, West Virginia, Ohio, Pennsylvania, Kentucky, Tennessee, Texas, Canada, Nova Scotia, New Brunswick and California all supply it in greater or less quantities. It occurs in nearly every geological formation, from the lower Silurian to the tertiary epoch. It is found associated with sandstones, limestones and shales, sometimes existing in subterranean cavities; sometimes saturating the porous rocks; sometimes impregnating the loose sands. It is met with at various depths, from a few feet to more than a thousand. Most of the productive wells in Pennsylvania are over 300 feet deep. The oil occurs in several zones at different depths, the higher zones furnishing the heavier oils and the lower zones the lighter. The fissures containing the oil usually occur upon gentle anticlinals, the oil and gas being forced up by the pressure of water from the synclinal troughs. The flow of oil is almost always accompanied by carburetted hydrogen gas, which rushes up sometimes with an almost irresistable force. Salt water also, especially in Pennsylvania, accompanies the oil. The proportion of salt water is very variable. The probable manner in which the water, oil and and gas occur in the subterranean cavities, is illustrated in the following diagram :

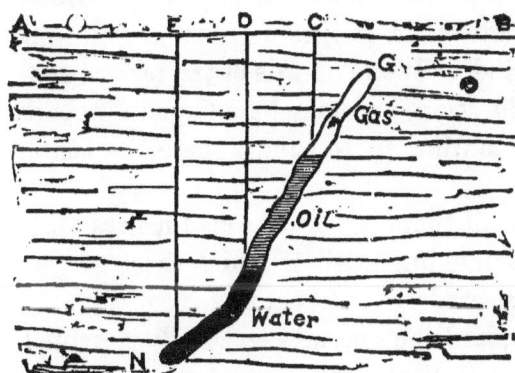

A B represents the surface of the ground. G N a fissure underground, filled with water, oil, and gas. A well sunk at C would give forth gas only. One sunk at D would give a flowing well, the oil being forced up by the supernatant gas, and the oil would continue to flow as long as the elastic pressure of the confined gas is greater than the weight of the column of atmosphere. The pressure of the gas is sometimes equal to several hundred pounds to the square inch; sufficient to throw the oil fifty or sixty feet high in the air above the surface. Should a well be sunk at E there would be a flow of salt water, and this would continue until the lower surface of the oil sinks to the level of the well, when oil would begin to flow.

Wells are sometimes intermittent and act precisely like intermittent springs. "More frequently, however," says Prof. Winchell, "the continual action of gas and oil produces the phenomenon. In boring a well suppose a stream of gas is struck one hundred feet from the surface of the rock and a small stream of oil twenty feet below the gas. The entrance of oil fills twenty feet of the hole and begins to submerge the fissure at which the gas is escaping. The gas forces its way through the oil with a spluttering sound, bubble after bubble rising to the surface. As the oil ascends the gas makes louder and louder complaints, till finally summoning all its accumulated energies it hoists the superincumbent column of oil to the surface, and pours it out in a stream of a few seconds' duration. The flow then ceases and the same operation begins to be repeated."

At other times the oil saturates a porous sandstone as water does a sponge, and when this sandstone is perforated by the auger the oil collects in it like water in a well sunk in a porous, humid sand. Wells of this character are generally of long years' duration, and may be pumped for years, the porous sandstone continuing to supply the oil as fast as pumped out.

As to the origin of oil there are numerous theories all of them more or less tenable, but to all of which there are objections not easily answered.

The most plausible, and the one now generally accepted, is that our coal beds, coal oil, bitumin, and such substances are the products of organic life, animal and vegetable, which flourished in the ancient Silurian, Devonian and carboniferous seas. The remains of this life are seen in the multitude of fossil shells, in the bunches of seaweeds, and in the extensive coal beds which occupy such a large portion of this carboniferous era. Nearly all the black shale of the Devonian age appears to be nothing more than compressed beds of seaweeds, which were deposited in the oozy bottoms of an ancient ocean. This hardened bed of seaweeds is the most prolific source of petroleum. The Cincinnati rocks, in their accumulations of animal remains, supply material for the production of coal oil. Exactly how the petroleum is extracted from these substances is not so clear, but it is supposed to be by a slow process of fermentation or distillation, but the fact remains that animal or vegetable remains are a necessary antecedent to the generation of coal oil.

Mr. H. Byassen explains the origin of petroleum, upon experimental grounds, as follows:

If a mixture of vapor of water, carbonic acid and sulphureted hydrogen be made to act upon iron heated to a white heat in an iron tube, a certain quantity of liquid carburets will be formed. This mixture of carburets is comparable to petroleum. The formation of petroleum can thus be naturally explained by the action of chemical forces. The water of the sea, penetrating into the terrestrial crust, carries with it numerous materials, and especially marine limestone. If the subterranean cavity permits these new products to penetrate to a depth where the temperature is sufficiently high, in contact with metallic substances, such as iron or its sulphurets, we have a formation of carburets. These bodies will form part of the gases whose expansive force causes earthquakes, volcanic eruptions, etc. Petroleum is always found in the neighborhood of volcanic regions or along mountain chains. In general it will be modified in its properties by causes acting after its formation, such as partial distillation, etc. Petroleum deposits will always be

accompanied by salt water or rock salt. Often, and especially where the deposit is among hard and compact
rocks, it will be accompanied by gas, such as hydrogen, sulphureted hydrogen, carbonic acid, etc.

The solution of the mystery of its formation is less important than determining the proper condition of its storage. When the rocks containing elements for generating
oil lie in juxtaposition to porous sandstones and limestones,
or are next to a stratum filled with cavities, the whole series being wrapped up in an impervious clay; the work of
generation, and the conditions of storage are found together, and such a region constitutes an oil region. Both
of these conditions must co-exist. The oil may be generated and lost, or there may be a fit receptacle for oil without the elements necessary to produce it. Observation has
shown that the best storehouse is a porous sandstone or conglomerate, and next to this a vesicular or cavernous limestone. Sometime good supplies have been found in the in-
terstices of hard siliceous rocks. This was the case with
the most productive wells in Overton County, Tennessee.

All petroleum oils are hydro-carbons, that is to say, they
are composed of the elements of carbon and hydrogen, and
range from light and inflammable oils, to those which are
heavy, viscid, and tarry, and requiring a high temperature
for ignition. According to Mr. S. Dana Hayes, State
Chemist of Massachusetts, petroleum yields by distillation
nine products, as follows:

Name.	Sp. grav. (Water, 1.)	Sp. grav. Baume.	Boiling Point.
Rhigolene	.625	—	65° Fahr.
Gasolene	.665	85	120° "
C. Naphtha	.706	70	180° "
B. Naphtha	.724	67	220° "
A. Naphtha	.742	65	300° "
Kerosene Oil	.804	45	350° "
Mineral Sperm Oil	.847	36	425° "
Neutral Lubricating Oil	.883	29	575° "
Paraffine	.848 (?)	—	—

Rhigolene is the lightest of all the products of petroleum,

evaporating rapidly in the open air and probably might be used in the manufacture of ice. It is now used for producing local anæsthesia in surgical operations.

Gasolene is used in automatic gas machines, and has been lately used in a stove, constructed for the purpose of cooking.

A., B. and C. Naphthas are used in paints and varnishes.

Kerosene is the oil used in lamps for illuminating purposes.

Mineral sperm oil is also used for illuminating purposes. It will not take fire at any temperature below 300 F. It is, therefore, very safe.

Neutral lubricating oil is the most valuable mineral oil for lubricating purposes. It has but little more taste or odor than the oil of almonds.

Paraffine is used for making candles, for preserving meats, to coat paper for photographic uses, to preserve timber, fruit, etc., to prevent oxidation of metals, to render fabrics waterproof, and for many other purposes.

The extent to which it has been brought into use during the past eighteen years, giving employment to the idle, light to darkness, heat to cold, health to disease, almost challenges our belief. Thousands of men, women and children find employment by its development. A writer connected with the Galena Oil Works, thus eloquently speaks of it:

"Petroleum has become an important article of commerce, requiring hundreds of vessels to transport it to the most distant lands. Kerosene, its refined product, is generally known all over the civilized world. It has found its way to every part of Europe and the remotest portions of Asia. It shines on the Western prairie, burns in the homes of New England, and illumines miles of princely warehouses in the great cities of America. Everywhere is it to be met with in the Levant and the Orient, in the hovel of the Russian peasant and the harem of the Turkish pasha. It

is the one article imported from the Uni ed States and
sold in the bazaars of Bagdad, the 'City of the Thousand
and one nights.' It lights the dwellings, the temples and
the mosques, amid the ruins of Babylon and Nineveh. It
is the light of Abraham's birthplace, and of the hoary city
of Damascus. It burns in the Grotto of the Nativity at
Bethlehem, in the Church of the Holy Sepulchre at Jeru-
salem, on the Acropolis of Athens and the plains of Troy,
and in cottage and palace along the banks of the Bospho-
rus, the Euphrates, the Tigris and the Golden Horn. It
has penetrated China and Japan, invaded the fastnesses of
Tartary, reached the wilds of Australia, and shed its ra-
diance over many a dark African waste. Pennsylvania
petroleum is the true cosmopolite, omnipresent and omnipo-
tent in fulfilling its grand mission of enlightening the
whole universe! Surely a product of nature that has become such a controlling influence in the affairs of men may
well challenge universal attention to its origin, its history
and its economic uses."

No oil ought to be used for household purposes which
can be set on fire, at an ordinary temperature, with a
match. The experiment can be tried by any one. Let
some of the oil be poured in a saucer; apply a match, if
the oil burns, it is unsafe and should be rejected.

The amount of oil obtained from some of the wells almost
defies credulity. The Empire well in Pennsylvania flowed
3,000 barrels daily for a year, when it ceased, and 200 barrels per day were brought up by pumps. The Noble well
made for its owners $11,000,000. The Phillips well yielded
2,000 barrels per day for a long time. The Sherman well
flowed at the rate of 1,500 barrels a day for some time,
then dropped to 700 barrels per day, and flowed steadily for
twenty-three months, and then become a pumping well. A
well sunk in Pennsylvania in May, 1877, flowed 3,500 barrels per day for three days, and then dropped down to 2,700
for two days, then 2,000 for a week; afterward 1,000 for

thirty days.   Two months after it was bored it only yielded
400 barrels per day.   The well seems not to be fed by per-
colation, but the oil appears to have been collected in an
underground cavern.

Dr. Winchell, in his Sketches of Creation, gives a list of
the flowing wells in Ontario, Canada, numbering thirty-
four, fifteen of which flowed over one thousand barrels per
day.   The well of Black & Matthewson discharged 7,500 barrels per day, and the oil floated on Black Creek to the depth of
six inches, and formed a film on the surface of Lake Erie.
Prof. Winchell says during the spring and summer of 1862
no less than five millions of barrels of oil floated off upon
the waters of Black Creek, a national fortune totally wasted.

It has been stated that petroleum occurs in almost every
formation.   The oil-producing region of Northwest Pennsylvania lies mainly outside of the coal field.   In South-
west Pennsylvania the wells are bored through some portions of the coal measures.   The oil strata pass below the
coal measures five or six hundred feet.   The conglomerate
which underlies the coal caps a few of the highest hills.
The oil wells are bored mainly in the Chemung and Portage
groups of the Devonian that lie just above the black shale
formation.   These two members of the Devonian (Chemung and Portage) are wanting in Tennessee, the only representative of the Devonian we have being the black shale
corresponding to the Hamilton shales of New York.   Prof.
Newberry considers this the mother rock of petroleum, containing much carbonaceous matter, which in nature's labo-
ratory is transformed into petroleum.   This is forced upward
into the porous rocks above by the water that finds its way
beneath, and by the pressure of the carbureted hydrogen
gas also furnished from the same material.   In Canada the
oil comes from the Hamilton shales.   Dr. T. Sterry Hunt
thinks the Corniferous limestone is probably the source of
petroleum in Ontario, but Prof. Winchell totally dissents
from this opinion, and declares that after having examined

all the oil regions east of the Rocky Mountains, he is convinced that the black shales are the chief generators of supplies of native petroleum. Below is a synopsis of the oil regions as given by Dr. Winchell, which goes a great way to sustain his position. It ought to be remarked, however, that the black shales around Burksville are precisely the same as those which occur in the oil region of Tennessee and belong to the Devonian age :

"I. The black shales of the Cincinnati group afford oil which accumulates in the fissured shaly limestones of the same group, and supplies the Burkesville region of Southern Kentucky, and Manitoulin Island in Lake Huron.

"II. The Marcellus shale affords most of the petroleum which accumulates in the fissured shaly limestones of the Hamilton group, and thus supplies the Ontario oil region, locally divided into the Bothwell district, the Oil-Springs district, and the Petrolea district.

"The Marcellus shale affords also a large portion of the oil which accumulates in the drift gravel of the Ontario region.

III. "The Genesee shale, with perhaps some contributions from the Marcellus shale, affords oil which accumulates in cavities and fissures within itself in some of the Glasgow region of Southern Kentucky.

"It affords also the oil which accumulates in the sandstones of the Portage and Chemung groups in Northwestern Pennsylvania and contiguous parts of Ohio.

"It affords also the oil which accumulates in the sandstones of the Waverley (Marshall) group, in Central Ohio.

"It affords also that which accumulates in the mountain limestone of the Glasgow region of Kentucky and contiguous parts of Tennessee, as also some of that which is found in the drift gravel of the Ontario region.

"IV. The shaly coals of the false Coal-measures, aided, perhaps, by the Genessee and Marcellus shales, seem to afford the oil which assembles in the coal conglomerate as worked in Southwestern Pennsylvania, West Virginia, Southern Ohio, and the contiguous but comparatively barren region of Paint Creek, in Kentucky.

"V. The Coal-measures may perhaps be regarded as affording a questionable amount of oil, which may have been found within the limits of the Coal-measures in the West Virginia and neighboring regions.

"From this exhibit it appears that the principal supplies of petroleum east of the Rocky Mountains have been generated in four different formations, accumulated in nine diferent formations, and worked in nine different districts."

Something about the Production in Pennsylvania.— Since the striking of the Drake well in the Pennsylvania oil region in 1859, there has been produced in that State up to January 1, 1877, 82,026,500 barrels of petroleum, which brought at the wells the sum of $287,000,000 to the producers.

In the year 1859 the production of the Western Pennsylvania region was only 3,200 barrels. This sold in its crude state at the wells for an average of 31 cents per gallon. The production for the following year (1860) aggregated about 650,000 barrels. The price for this year had diminished to such an extent that the crude oil sold for only 16 cents per gallon.

Subsequently the highest price realized for oil was obtained in 1864, when the aggregate production reached 2,116,182 barrels. The average price received for oil during this year was $7.62 per barrel. In 1874 the production reached the enormous sum of 10,910,303 barrels, which had a great effect upon the price of oil for that year, the average price for the year being only $1.29 per barrel. Since that time the production has gradually fallen off. Until quite recently serious apprehensions were entertained of the final exhaustion of this important commodity in the Pennsylvania oil region. Under the effects of this diminution in the amount of production the price of crude petroleum at the wells rose to $2.73 per barrel. This, however, has been effected by the discovery of large producing wells in the Bullion District, the extent of whose production has proven almost as great as in the best periods in the history of oil interest in this country. Since April of this year the price of this substance has ranged from $2.40a$2.93 to $1.50 per barrel in June.

It is estimated, by competent persons, that the cost of refining the oil produced for a succession of fifteen years has been about 75 per cent of the cost of the crude material.

From 1859 to January 1, 1869, there had been 5,560 wells drilled, which produced something less than 25,700,-000 barrels of oil, making an average yield for each well of nearly 4,600 barrels. For this was realized an average of $4.06 per barrel, making the average sum for each well $18,700.

From 1869 to the present time the nature of the producing territory has been better understood than previously, and as a consequence there has not been so great a number of failures in drilling. The total number of wells drilled from 1st January, 1869, to 1874, was 4,939, each of which produced about 8,400 barrels. The average price of crude oil for this period is $2.91, making the product of each well yield about $24,500. Since that period up to April, 1877, 8,902 wells have been sunk.

At the last of April, 1877, of the 17,955 wells which had been sunk on or near the producing territory, 6,846 were pumping, producing an average of five barrels each per day. The daily average of new wells is thirteen and a half barrels. The average life of wells in the Pennsylvania region, taken from actual record, is a little over two and one-half years. The cost of drilling the 5,560 wells up to January, 1869, was $4,000 each. For the remaining wells during the latter period mentioned, no estimate has been made.

About one well in six sunk in the oil region proves a dry hole. Many others do not yield over four or five barrels per day at first, with a constantly diminishing return.

The following tables will show some interesting facts in regard to petroleum. They are taken from *Stowell's Petroleum Reporter*:

PRODUCTION.

AMOUNT OF PRODUCTION EACH MONTH SINCE JANUARY 1st A. D.

A.D	January	Febru'ry	March	April	May	June	July	August	Septem'r	October	Nov'ber	Dece'ber
1870	11,287	11,714	11,299	12.294	13,211	13,731	13,735	16,776	18,462	18,171	16,953	15,178
1871	13,497	13,305	12.914	12,866	13,187	13,678	14,725	14,622	15,398	15,653	15,487	15,418
1872	18,825	15,965	14,890	15,403	17,326	16,371	16,702	17,739	16,681	4,272	21,287	20,825
1873	20,407	21,725	21,467	21,384	25,044	26,449	27,983	30,198	31,809	30,403	33,049	34,980
1874	37,653	29,839	28,498	25,958	28,895	30,725	33,337	30,049	28,021	29,669	28,702	27,682
1875	27,489	25,708	25,469	22,502	22,468	23,207	25,431	23,186	23,298	23,583	23,340	23,254
1876	22,975	23,065	23,167	23,383	23,721	24,120	24,633	235,53	26,020	26,102	26,216	25,390
1877	27,190	27,979	29,087	32,427								

STOCK.

TOTAL STOCK OF CRUDE in the OIL PRODUCING FIELDS EACH MONTH SINCE JAN. 1st, A.D.

A.D.	January	Febru'ry	March	April	May	June	July	August	Septem'r	October	Nov'ber	Dece'ber
1870	340,158	352,390	351,474	328,609	329,908	351,168	321,840	356,908	419,477	473,896	576,014	554,629
1871	537,751	587,021	642,040	771,000	605,000	554,000	511,220	530,146	541,300	495,102	502,960	532,000
1872	532,971	579,793	662,497	877,832	950,603	1,010,302	990,229	997,196	951,410	914,423	886,909	1,084,423
1873	1,183,728	1,265,373	1,244,657	1,178,643	1,192,541	1,324,493	1,433,620	1,513,890	1,521,185	1,452,777	1,493,875	1,625,157
1874	1,948,919	2,283,032	2,648,210	2,623,534	2,594,286	2,701,625	2,279,479	2,932,444	2,758,504	3,134,902	3,449,845	3,705,639
1875	4,104,703	4,496,751	4,592,364	4,537,843	4,552,672	4,502,896	4,386,720	4,223,397	3,812,945	3,072,101	3,701,235	3,555,200
1876	3,585,143	3,734,835	3,829,250	3,900,703	3,989,904	3,791,642	3,326,726	3,304,405	2,930,456	3,040,108	2,955,092	2,551,199
1877	2,604,128	2,860,636	3,210,454	3,279,731								

Average price per barrel of 42 gallons for the crude oil at the wells:

1859	$20 00	1869	$5 48
1860	9 60	1870	3 74
1861	49	1871	4 50
1862	1 05	1872	3 84
1863	3 15	1873	1 84
1864	7 63	1874	1 29
1865	5 18	1875	1 48
1866	3 78	1876	2 73
1867	2 54	1877	2 87
1868	3 05		

The effect of increased production will have a tendency to lower prices, but the decline in prices will bring it into new uses. The time is not distant, when some of the products of petroleum will be used in heating, as well as lighting our houses, cooking our food, and in various other ways ministering to the comforts of life. The New York correspondent if the *Titusville Herald*, commenting upon this increased production, says: Europe must have the oil, and refiners are going to make them pay for it. Europe is trying to buy without agitating the market. The present disparity between production and consumption, and the excess of the former over the latter, will avail European merchants nothing during the present season.

This disparity is well understood at the seaboard and in Europe, and it seems pretty well understood that Europe is to have no benefit from it. There is plenty of empty tankage in the region, and into tanks the surplus will go, if there shall be any surplus, although it is doubtful if there will be any, at least for some months to come. What, though production now reaches 30,000 barrels per day, with a prospect of further increase to 32,000 by the first of July? The actual amount of accidental consumption last year averaged 2,000 barrels per day during the first six months and it is only reasonable to suppose the lightning will carry off as much during the present year. The domestic trade, which is not over well supplied, will require say 6,000 barrels per day which will increase with the advent of warm weather when the demand for refined oil for petroleum stoves will commence. And it very evident that the enormous demand of the export trade in 1876 will be materially increased during 1877. During last year nearly 10,000,000 barrels were shipped from the region and it is universally conceded that these shipments did not represent the actual consumption of the world, the surplus stocks abroad at the beginning of the year having been drawn upon to supply the deficiency. The present foreign stocks being much smaller than a year ago, and affording a smaller reservoir to draw upon will necessitate larger shipments to meet demands for cousumption.

CHAPTER II.

OIL REGION OF TENNESSEE.

GEOLOGY—TOPOGRAPHY — SOILS — STREAMS—FARM PRO-
DUCTS—TIMBER, OF OVERTON, PUTNAM, CLAY, JACKSON
AND FENTRESS COUNTIES.

The oil territory of Tennessee occupies the extreme southern end of the great oil belt which extends in a south-west-
erly direction from Ontario, Canada, through New York,
Pennsylvania, West Virginia, Kentucky and Tennessee.
This belt widens out at both extremities, leaning to the
westward in Canada, and spreading out laterally in Tennessee, so as to include Dickson and Hickman counties.
The following counties are included, or supposed to be
included, in the oil region of Tennessee, viz.: Overton,
Clay, Putnam, White, Warren, Jackson, Trousdale, Sumner, Davidson, Dickson and Hickman. In all these indications of petroleum have been found. Overton, Clay,
Putnam, Jackson and Fentress furnish by far the most
numerous external indications of oil, and to a particular
description of these we shall address ourselves.

Geology.—These counties mainly belong to that natural
division of the State called the Highland Rim, a portion of
the State which surrounds, like the rim of a dish, a great
central basin in the interior of the State. For the most
part the formation is sub-carboniferous, though the deep
bedded streams usually cut down through this formation
and the Devonian black shale to the Cincinnati or Nashville group. The Devonian age, unlike that in Pennsylva-

nia has only one representative corresponding with the
Hamilton black shale of New York. The Corniferous,
Marcellus, Portage, Chemung and Catskill are all wanting,
also the upper Silurian rocks. The Portage group in New
York and Pennsylvania, consists of shales and laminated
sandstones, and has a thickness of a thousand feet or more.
The Chemung group has the same lithological character,
and in these two formations, most of the oleaginous product
of Pennsylvania as has been mentioned, is stored.

OVERTON COUNTY.

Geological Formations.—In Overton county there is a
great variety of formations. The coal measures cap the high
points in the eastern part of the county, and are also met
with in isolated peaks, as Pilot Mountain and Alpine Mountain. The Lower Carboniferous consisting of two groups,
the Mountain limestone and the Siliceous group, has a wide
spread development, especially the latter. The Mountain
limestone occurs in limited areas, forming benches on the
slopes of the Cumberland table-land, and crops out on the
sides of extensive plateaus which rise above the general level
of the county. It is about 400 feet thick.

The Siliceous group covers fully three-fourths of the county,
and forms the clay uplands of the county. This group has
two members, 1. The Lithostrotion or Coral bed. 2. The
Lower Siliceous or Protean bed, lying below.

The Coral bed covers extensive areas in the northern parts
of the county, and may always be known by the presence
of a fossil coral resembling a " petrified hornet's nest." It
is usually about 200 feet thick in this county.

The Lower or Protean bed covers all that region around
Spring creek, the undulating lands to the south of Livingston, the flat lands on the west, and many places in the northwestern parts of the county. It takes numerous forms. At
the oil wells on Spring creek, it is a hard, flinty limestone ;
below Waterloo falls, and at places on Obey's river it is a

gray crinoidal limestone ; on Eagle creek, a bluish shale
known as the Keokuk shale, and in other localities a siliceous rock. Its general thickness is about 270 feet.

Below this comes the black shale of the Devonian age,
varying in thickness from 26 to 35 feet. It is seldom found
as a top formation. A few limited areas are found at the
mouth of Eagle creek, and in the beds of some of the streams
as they near Obey's or Cumberland rivers. Nevertheless, it is a very persistent formation, and may always be
found in its proper horizon.

The Cincinnati group of the Lower Silurian presents
itself in the beds, or enclosing banks, of the principal streams
as they approach the Cumberland. On Spring creek the Cincinnati or Nashville rocks come to the surface below Waterloo falls ; on Roaring river, just below Crawford's mill.
On Obey's river these rocks begin just above the mouth of
Franklin creek.

Topography, Soils and Streams.—As might be inferred
from the great diversity of geological formations, the topographical features are striking, and show variety in an emi-
nent degree. There are rugged heights, rolling plains, level
plateaus, rocky gorges and deep-sunk valleys, and coves
sheltered between the massive walls of everlasting hills.
The Cumberland table-land has scolloped edges displaying a
very rugged contour, steep escarpments, sloping sides rough-
ened by boulders. Level areas on the summit are characteristic of this portion of the county. The sides of the
mountain are furrowed by many a stream. Chasms great
and terrible, profound in their depths and striking in their
suddenness, form one of the principal features of the western mountain side. Spurs shoot out for many a mile into
the lower plains. Many of these are dissected by transverse
hollows, leaving isolated peaks nearly or quite as high as
the Cumberland mountain itself. The top of the mountain
has all the characteristics peculiar to this division of the
State. A thin soil resting upon a conglomerate sandstone,

from the crumbling down of which it has been derived, everywhere marks this portion of the county. Two or three seams of good coal are usually found under this cap of conglomerate rocks, inter-stratified with shales and sandstones. The soil of the mountain is adapted to the growth of apples, Irish potatoes and garden vegetables, generally. The land is thinly wooded, covered usually in sumner with a luxuriant growth of native grasses and pea-vines, and furnishing a large amount of highway pasturage. But few habitations are found here, and this part of the oil region is almost as wild as it was when the Indian roamed in all his fearless independence through its silent forests. But little of the table-land, however, belongs to Overton or Putnam counties. An outlying ridge in Overton county extends northward between East and West Fork of Obey's river, about fifteen miles in length, and in breadth from one to six miles. The outlines of this ridge are very irregular. At two or three points it is broken by gaps which drop down to a level with the terrace-lands which lie all along the western borders of the Cumberland table-land. The ridge between the East and West Forks (see map) often rises to the height of the table-land, that is to say, about 2,000 feet above the sea, and 500 feet above the average level of the terrace-lands.

These terrace lands occupy a considerable part of the county. From below the cliffs of the table-land they extend westward and northward, oftentimes indented by coves or cut by streams, along which lie many beautiful and fertile valleys. It is impossible to define with accuracy the limits of this portion of the county. It occupies all the south-eastern part of the county except a small strip that lies in the Cumberland table-land. On a level with these terrace-lands there is a ridge covered with sandstone extending through the middle of the county, forming the watershed between the waters of West Fork and Roaring river. It extends from Thorn gap thirteen miles south-east of Livingston to a point near Monroe, a distance of about sixteen miles. It then turns to the westward, widening out into terrace lands and extending about fifteen miles further. The breadth at some points is five or six miles, while at others it is nearly cut in two by coves on opposite sides. Usually the top of this ridge is a well defined plateau, but at several points there are knobs rising from its surface which attain an elevation nearly equal to the table-land. Pilot Mountain is one of these, which lies a little south of Monroe, and is really an outlier of the table-land with its seams of coal, and the characteristic sandstone of that division of the State. Alpine mountain is another, a little east of Livingston. The soil of these terrace-lands is sandstone, and much resembles that of the table-land. Indeed, it was probably a portion of the table-land, but the erosion of time has gradually carried away the coal measures, except at a few places, and left the ancient foundation upon which they rested. In traveling from Livingston to Celina, the county seat of Clay county, the road ascends the terrace-lands a little northwest of Livingston, and keeps on it to within a few miles of Celina. The underlying rock is mainly the Mountain limestone, and some of the soil is very productive.

The valley of West Fork lies between the ridge which runs from Thorn gap northward, and the outlying ridge between East and West Fork. The usual breadth of this valley is about two miles, and its length about twenty. It is one of the best portions of the county for agricultural purposes. In the angle formed by this dividing ridge in its turn westward, there is a beautiful little valley, in which Livingston is situated.

Numerous coves exist in the neighborhood of Livingston that supply much fine farming lands. White's cove begins at Livingston and extends eastward for three miles. It is bounded by high plateau lands on every side except the east. It is rather an indentation in the great dividing ridge or watershed that we have spoken of. A considerable quantity of land in White's cove is swampy. The native growth is sweet-gum, swamp-ash and white-oak. The slopes bounding this cove are very heavily timbered with poplar, chestnut and black walnut. The soil of the slopes though very fertile, is rendered almost unfit for agricultural purposes, by reason of large masses of sandstone and limestone, which everywhere cover the surface. Going eastward, to the head of the cove, and ascending the terrace-lands, we find them level and heavily timbered. Alpine mountain rises like a truncated cone from these terrace-lands, and forms a striking object in the landscape. The top of Alpine mountain is level, but a deep gorge called Medlock's hollow, about two miles long, cuts the mountain towards the north-east into two arms. On the sides of the gorge coal seams are exposed, but they are thin and nearly worthless. In rock houses on the face of the escarpment, alum and copperas are found in great quantities, resulting from the disintegration of the aluminous shales. Transverse gorges make into Medlock's hollow, but they are usually short. Brown hematite iron ore occurs on the side of Medlock's hollow in large quantities, but it is very siliceous. Chalybeate springs abound. Upon this mountain there is every condition of health. There is a life-giving property in the atmosphere that imparts an elasticity to the frame, giving joy to the heart and an animation to the soul. The mind and body receive a new vitality by being bathed in this pure mountain air, which produces an exhilaration of spirits beyond that of any drugs. The time is not now, but it will come, when the edges of this mountain will be covered with the palaces of the rich, who will come to enjoy the perpetual delight of breathing the mountain air.

Bates' cove lies three miles north of Livingston, and is scooped out of the terrace-lands. It is two and a half miles long and two miles wide. It furnishes much good farming land, and the farms are usually well improved.

Dog Hollow cove lies thirteen miles south of Livingston,

on the headwaters of West Fork, and furnishes the largest
and best body of farming land in the county.

Big Joe Copeland cove lies four miles south of Livingston, and is about three miles long and three-fourths of a
mile wide. The soil is excellent, and was originally covered
with cane.

Hunter's cove lies to the north-west of Livingston, and
includes 600 acres of very fertile land.

Numerous other coves, as Eldridge's, Hunter's, Nettler's,
Sinking cove, Andrew's cove and Carlin's, supply a large
amount of very productive lands. In fact, the existence of
rich coves form one of the chief attractions of Overton
county. Many of these coves furnish as good soils as may
be found in the State, which will produce year after year
large crops of grain and forage, with scarcely any perceptible deterioration. The soil is a good loam and easy of
tillage.

The lands in the south-western part of the county are
much broken, except on the plateaus between the streams.
Roaring river, with its tributaries, (Spring creek being the
largest,) drains this part of the county. The slopes to the
streams are very gradual, and the alluvial bottoms narrow.
The plateaus between the streams furnish some soils of a
moderate fertility and a large amount of valuable timber.
These plateaus are what are known as clay uplands. This
class of lands occupy most of the area of the valleys also.
The surface is moderately undulating, sometimes rolling or
broken and easily washed. The Lithostrotion bed covers
much of the county lying below the terrace-lands. Frequent sinkholes are seen. The St. Louis limestone is seen
on the hill-sides and along the streams near their sources,
but lower down the streams cut through it; also through the
black shale below, to the Nashville rocks. But little of the
soil, however, rests upon the Nashville limestone, only a few
bottoms on Obey's river, Roaring river and Mill creek. In
the northern part of the county the ground is strewn with a

fossil coral known as the Lithostrotion Canadense. In many places a coarse black chert is also scattered over the surface. This invariably denotes a good soil. Upon some of the hills in the northern part of the county, notably so in those bounding Eagle creek and Obey's river, a fine-grained sandstone occurs, known as the Waverly sandstone. It furnishes a poor soil. Toward the western part of the county frequent masses of coarse sandstone are met with. Near Hillham the forests are very dense. Chestnut timber abounds. The soil, when fresh, produces well, but is easily exhausted unless rotated with clover.

The soils of the county may be put in the following classes:

1. *Cumberland mountain sandstone soil.*—Thin and unproductive for ordinary field crops. Good for fruits and vegetables, especially Irish potatoes.

2. *The soil of the terrace-lands.*—Composed mostly of sandy material, but with more humus than the soil of the mountain, and consequently more productive. It covers an extensive plateau between Livingston and Celina. It forms the terrace lands of the county.

3. *Mountain limestone soil.*—Found on the slopes of the Cumberland table-land and on the terrace slopes. This is generally very productive but hard of cultivation, owing to the prevalence of surface rocks. Many of the upper benches of the mountain has a soil derived from this limestone, which is rich in plant food. It is highly productive of the cereals.

4. *The soil of the St. Louis limestone.*—This is distributed over a considerable portion of the northern part of the county. All that extensive plateau lying between West Fork and Eagle creek, and extending southward to Alpine mountain, is covered with this soil. Also a considerable area around Hillham. Its limits are precisely those of the coral bed spoken of on a previous page. This soil is clayey and filled with fossil coral. Some areas are very fertile. It

constitutes a large part of the clayey uplands. It often runs into the

5. *Siliceo-calcareous soil*—which lies below, but separated from it in places, by a fine-grained sandstone seen on the edges of the bluffs on West Fork, Eagle creek and Obey's river. The siliceo-calcareous soil abounds south of Livingston, extending as far up as the oil wells on Spring creek. It has cherty or flinty layers with a red retentive clay. The large black-oak forests south-west of Livingston, grow upon this soil, and much of the heavily timbered lands about Hillham belong to it. It is susceptible of indefinite improvement, the underclay retaining for a great while all fertilizers put upon it. Stiffer than other calcareous soils, it is not so liable to wash where the land is moderately broken. It has commingled with it beds [of chert which supply a natural drainage. Such lands should produce large crops of wheat and tobacco. The lands have the same soil upon which the heavy crops of fine tobacco are grown in Montgomery, Robertson and Stewart counties, and which extend into Christian, Todd, Logan and Simpson counties, of Kentucky. These soils are noted also for their durability and strength, and for the certainty with which they produce crops. Droughts effect them less than any other soils in the State, when they have been well and deeply plowed, the clay below acting as a reservoir which supplies the moisture to the plant during the dryest seasons.

The siliceous character of this soil is lost in some of the lands on West Fork, where the siliceous limestones give way to the Keokuk shales. The soil from these shales is less friable, being stiff and cold and of an ashen hue. It is not so productive as the siliceous soils, but this want of productiveness is not caused so much by a lack of plant-food as by its bad physical condition. It is so compact and stiff that free access of the air is prevented, which is essential to the fertility of soils and to the healthy growth of plants. Subsoiling and under-draining would make the soils from the Keokuk shales as valuable as any in the county.

6. *Black shale soil.*—This is confined to a few farms near the mouth of Eagle creek. It is stiff but forms good grass lands, and when deeply stirred by the plow produces a bounteous yield of corn and wheat.

7. *Nashville limestone soil.*—This is very fertile, probably the strongest soil in the county, but limited in area, being confined to a few bottoms on Obey's river, Mill creek and Roaring river.

It will thus be seen that Overton county, and indeed all the counties in the oil belt, possess a great variety of soils, and are adapted to the production of every crop east of the Tennessee river.

Fruit, owing to this great variety of soil and to the endless variety of exposures and sub-climate arising from differences of altitude, is never a complete failure.

Overton county is abundantly supplied with streams which ramify every square mile of its surface except a small portion of the terrace-lands. Obey's river is the largest and most important stream, though it only passes through a small portion of the county on the north-east. It is navigable as far as the mouth of West fork, for several months in the year. West Fork, its main tributary, rises twelve miles south-east of Livingston, and flows in a northerly direction through a beautiful and fertile valley. Throughout its whole course it is a bold, rapid stream, hemmed in by high banks, and is capable of supplying a large amount of valuable water power. It has three forks: The most eastern heads in Norrid's cove in Overton county; the middle in a depression called Pine Hollow; the western in Dry Hollow.

East Fork rises among the mountain spurs of Fentress and Putnam counties, and by its union with West Fork, which is just within Clay county, forms Obey's river. It much resembles West Fork, though longer and larger. The two streams head within a short distance of each other and enclose the high ridge heretofore spoken of in their circling arms. Most of this stream is in Fentress county. Some excellent seams of coal are exposed by its head branches. It is also found at many places along its course which will be spoken of hereafter. Roaring river, a tributary of the Cumberland river, ramifies with its head-waters the county south of Livingston and flows' north-west. It is a bold, impetuous stream, and supplies some excellent water powers. The most important tributary of this stream is Spring creek, which rises in the mountain spurs, and, taking a north-westerly course, forms for a considerable distance the boundary between Putnam and Overton counties. On the banks of this stream the largest quantities of oil have been found. A more minute description of the operations there will be given hereafter.

WATERLOO FALLS

on this stream, is a noted water power, and deserve a full description. These Falls are ten miles south-west of Livingston and ten miles north of Cookville, the county seat of Putnam county. They consist of a succession of three falls. The bluff above the first fall is 96 feet high, but is easy of descent by a winding road. The first fall tumbles over ledges of siliceous limestone eight feet thick. From the foot of the first fall, for 800 feet, the descent of the stream is very rapid, falling at the rate of about one foot in fifteen. On the right of the stream which here sweeps around in a semi-circle, is a level space covering several acres, affording ample room for the erection of any number of mills. On the left a precipitous bluff of siliceous sub-carboniferous limestone rises to the height of 100 feet or more, forming a bounding circular wall. The water from the first fall can be easily and cheaply conveyed to the point indicated, and its whole force turned upon a water-wheel A powder mill was at one time in operation at this place. Four hundred and fifty feet below this point another fall occurs of about two feet. This is succeeded, at the distance of one hundred and twenty feet, by the grandest of all the falls on the stream. Here the water has a perpendicular descent of 10 feet, striking at that distance against a mass of shelving rocks which forms a bench. The water as it falls upon these rocks is dashed into spray, and rolls over the projecting layers of rocks in great white foaming masses, into a quiet pool below, where, resting awhile, it darts off into a succession of cascades. The top rock of the falls is a hard siliceous limestone about three feet thick. This rests upon the black shale, which is here about 24 feet thick. The stream below the main falls flows over the black shale for a hundred yards, at the end of which distance it cuts down into the grayish Nashville limestone. The gorge widens and deepens below the falls, the rocks of the bluff sometimes overhanging the yawning chasm. The black shale is saturated with petroleum, and even the Nashville rocks below are made odorous with the oleaginous exudations. The stream comes around, after going half a mile, to within 700 feet of the point of deflection above the falls, forming a peninsula. A race cut across the neck of the peninsula will secure a fall of 40 feet. The average width of the stream is 40 feet, and depth about fifteen inches. The water is constant, being fed by never-failing springs. The banks of the stream are durable, and the material for the construction of dams is abundant. The neck of the peninsula through which a race may be cut, consists of a tough clay with intermingled chert and gravel. I know of no more valuable water power in the State. The only disadvantage attending it is the distance from means of transportation.

A bed of crinoidal limestone probably 100 feet high, takes the place of the lower member of the siliceous group just below the falls. It rests upon the black shale, and is much eroded about the centre. The debris or talus which extends from the deep groove in the bluff down to the black shale formations, is composed in large part of crinoidal stones and buttons, some of them very large and beautiful.

Flat creek rises four miles north-west of Livingston, flows south-west, and empties into Roaring river. Its length in a direct line is four miles.

Mill creek rises north-west of Livingston, and takes a general direction north-west, to the Cumberland river, carving out as it approaches the river, a basin, deep down into the Nashville rocks. It has a few good bottom farms near its mouth.

Mitchell's creek heads eight miles nearly north of Livingston, runs north-west, and empties into Obey's river. It also affords good water powers. It has several falls. Near its mouth are some good bottom farms, resting upon the Nashville rocks. Carter's creek empties into this from the west side.

In going up Obey's river Irons creek comes next. It rises eleven miles north-west of Livingston, and empties into Obey's river two miles below the mouth of Wolf, which comes from the opposite side.

A mile above the mouth of Wolf, but on the south side of Obey's river, Asburn's creek comes in. It is a small stream, but noticeable from the oil seeps and salt wells that occur on its banks.

Eagle creek rises five miles north of Livingston, and flows north and north-east, and empties into Obey's river seven or eight miles in a straight line from the junction of East and West Forks. Near the source of this stream are some excellent water powers that have been utilized.

Several of these streams pass through a portion of Clay county, as may be seen by reference to the accompanying map.

PUTNAM COUNTY.

Having thus given a general view of Overton county, we turn now to Putnam, a county in many respects similar.

It lies on the south of Overton, and, geologically, is almost identical with Overton.

Geology and Topography.—Much of this county belongs, geologically, to the Lithostrotion bed of the siliceous group of the sub-carboniferous. The St. Louis limestone occurs in a strip on the west, running north and south. East of Cookville this is super-imposed by the Mountain limestone which forms many ridges. Still east of this is a strip very irregular in outline, of the carboniferous formation. A considerable ridge runs through the county from west to east, upon which the old Walton road passes. This ridge is the watershed between the waters of Caney Fork on the south and those of the Cumberland river on the north. The length of the county from east to west, is about forty miles, while its average breadth is not more than twelve miles. To go more into detail, the eastern end comprising about one-eighth of its entire area, is on the Cumberland table-land. This part of the table-land is remarkable as containing the head springs of streams radiating from it as a centre toward every point of the compass. The East and West Forks of Obey's river flow north, Spring creek northwest, Fallingwater nearly west, Calf Killer river southwest, and just across the line, in Cumberland county, are the head springs of Emory, which flows east into Clinch river, above Kingston. These facts are an evidence of its great elevation. These streams, except the last, in their descent from the elevated plateau, have cut through the western escarpment, forming many deep ravines and sequestered valleys, with towering ridges projecting between. The scenery here is remarkable for its wildness and sublimity. Bold cliffs of sandstone and conglomerate, crowned with scraggy trees, where the scream of the eagle is not infrequent, and the howl of the wolf is sometimes heard; mountain sides rugged with jutting cliffs, the deformities of which are sometimes concealed by mantling ivy; "benches" (terraces) here and there with good farms and orchards; deep

3

valleys sometimes with narrow bottoms, but more frequently pressing close upon a stream which dashes and thunders down one cascade after another—such are the characteristic features of this part of the county. As we approach the central part of the county, the valleys become wider, and the ridges and spurs run out into lower hills, or disappear entirely. We are now in the red clay region, a broad belt of which extends along the western base of the table-land. In Putnam county this belt is about fourteen miles wide, and is the best part of the county. Its surface is diversified with hill and dale, the beds of most of the streams being considerably below the general level of the country. Sinkholes and caves are a characteristic of this belt of clay lands, and in the neighborhood of the mountains are many large springs, whose waters have accumulated, and perhaps flowed for miles in underground channels. The country becomes more level and the lands less fertile toward the west, until the part of the county designated by the significant name of "barrens" is reached. Here the red clay gives place to a yellowish sub-soil, greatly deficient in calcareous matter, and too leachy to bear improvement. There is but little humus in the surface soil, and it is not well adapted to the production of grain. The surface is generally level, except in the neighborhood of the streams, and the timber is thin and of small size. But the valleys and the hill-sides along the streams afford some good lands, and the less fertile portions are covered with nutritious wild grasses, which furnish pasturage for large numbers of cattle and sheep. The extreme western end of the county runs down into the hills bordering the Caney Fork and Cumberland rivers, and takes in a small part of the Central Basin. The Highland Rim is so broken by the valleys separated by projecting ridges that its escarpment is not well defined. The surface is broken, but the soil of the valleys lying upon Silurian limestones is very fertile.

Soils.—The soils of the table-land are light and sandy,

and not valuable, except for fruit-growing and grazing. But little of this part of the county has been improved, and lands can be bought at very low figures. The mountain limestone on the western face of the table-land does not present any very extended areas of land level enough to cultivate, but there are several farms on some of the benches, which are rich enough to produce any crops grown in this latitude, and are especially valuable for fruit farms. In such situations orchards never fail to bear good crops. The cove lands are often level and always very fertile. The soil is a yellow loam, having enough of sand to render cultivation easy, but not so much as to impair its fertility. It is sometimes several feet thick, resting upon red clay or limestone. As already stated, the clay uplands occupy the central part of the county, and embrace the largest area of good lands. The soil is a dark brown mould, rich in humus, and with good tillage will continue to increase in fertility. The subsoil is a strong red clay, possessing many of the elements of fertility. At a greater or less depth beneath the surface is found limestone, either blue or gray, and sometimes fossiliferous. It often crops out on the hill-sides, and nearly always along the streams. The soils in the barrens are chiefly valuable for grazing. We believe there is no part of the State better adapted to the rearing of sheep. The coarse native grasses are nutritious, and the cultivated grasses grow finely. But the porous yellow subsoil is so leachy that we do not recommend these lands for grain farms. There are places, however, where red clay and limestone are found, and in all such the lands are rich.

It may here be mentioned that much of this land which has been considered very infertile is gradually undergoing a change for the better. The early settlers had a very pernicious habit of burning the leaves from the "barrens" every spring in order to facilitate the early growth of the native grasses upon which they relied for pasturage for their stock. This practice was kept up by their descendants, and the result was: 1. That no humus could accumulate, and 2nd, the shrubs and young trees were killed as fast as they appeared. The woods were therefore thin, and the trees consisted mainly of black-jack oaks, which were able, by their rough, thick bark, to resist these annual conflagrations. The land was, therefore, exposed to the heats of the summer sun. All moisture was dissipated, and no fertilizing gases could be absorbed from the atmosphere. Each succeeding year showed a greater degree of impoverishment. More recently, however, these fires have been unpopular, because they endangered the fences of the farms. They are, therefore, not so common. Wherever they have been kept out, the undergrowth has sprung up, and the land is shaded and kept moist during the summer. The soil is becoming blacker and more fertile, and the time will doubtless come when these lands will be in demand for tobacco and wheat, to the growing of which they are becoming well adapted.

The bottoms alongside the streams and hill-sides, especially those facing the north, are generally fertile. The valleys in the western end of the county have a deep, dark soil, generally resting on Silurian limestone, and very rich. Buffalo Valley, in the western end of the county, is four miles long, with an average width of one mile. The surface is level, and the soil very fertile. Dry Valley is scarcely less fertile, and has a large area. Along the base of the mountain are several coves, or small valleys.

Streams.—As has been stated the ridge running east and west is the water-shed of the county It is about 30 miles long. Beginning on the north side of the ridge there are several of the tributaries of the Roaring river, including Blackburn's Fork and Spring creek. On the south side is Falling Water, whose head-waters are within three miles of those of Spring creek, which flows in the opposite direction. The tributaries of Falling Water are Hudgens creek, Post Oak, Pigeon Roost, and Cane creek. West of

Falling Water is Mine Lick, which empties into Caney Fork. These last have a general direction south. Wolf creek, Indian creek and Rock Spring creek, run westwardly, emptying into the Caney Fork. The rapidity of all the streams of the county increases towards their mouths. Waterfalls characterize nearly all of them, the falls always occurring where the streams cut down through the black shale, which being soft and easy of disintegration soon gives way under the force of the water, while the hard siliceous rock lying above resists erosion for a great while. Many of these waterfalls, especially in White and DeKalb counties, are beautiful.

CLAY COUNTY.

This county, next to Overton, shows the greatest number of oil indications. It is about 30 miles in length, with an average breadth of twelve miles.

Geology and Topography.—These features of the county are easily described. Imagine a plain with a rolling surface, nearly level in the west. Imagine this plane cut from northeast to southwest by a deep valley a half mile wide and four hundred feet deep. This would represent the Valley of the Cumberland. Opening into the Cumberland Valley on the eastern side is the long, winding valley of Obey's river, with a general direction from east to west. A number of smaller streams with valleys of their own, extending outward at nearly right angles to the river, with their ramifications, cut up this plane into innumerable winding spurs, pointing to the river like the teeth of a saw. These ramifications of valleys run back among the highlands, gently rising toward their source. These valleys are always skirted by hills, ranging from 100 to 300 feet in height. Many of the creeks are short and enclosed always by high bordering ridges. So numerous are the breaks that some have doubted whether they have been produced by erosion, but think they mark the courses of the great cracks formed in the underlying rocks as they were hardening or shrinking or yielding to the forces of subsidence and upheavals during the unstable period of geological eras. With this opinion I cannot agree. The rocks are approximately horizontal. A few gentle curves are met with here and there, forming anticlinals and synclinals, but at no place do these depressions correspond with the profoundly eroded surface. Indeed, it often occurs that no disturbance of strata is exhibited where the gorges are deepest. At some few places there are signs of considerable disturbance. Near the mouth of Ashburn's creek the strata all dip northwardly at an angle of some five degrees, and fissures appear in the Cincinnati limestone from which petroleum issues. The point of the greatest disturbance with which I have met in Middle Tennessee is in Jackson county, near the head of Flynn's creek. Here the rocks are tilted so as to be almost perpendicular, and still further down the creek there is a drop of 200 feet, which covers probably 50 acres. The black shale on the south side of the creek is elevated 200 feet above it on the north side, where it comes down and the water of the creek runs over it for half a mile to old Antioch Church. A short distance from this it appears on the side of the hill in its original geological horizon.

Beginning at the heads of these valleys, we find the lower carboniferous rocks, including the St. Louis limestone and the barren group of the Siliceous formation. These two groups are about 400 feet thick. Below comes the black shale, 25 feet thick, and then the Cincinnati or Nashville rocks of the lower Silurian, through which the Cumberland has cut about 80 feet at Celina. At Celina the black shale is 78 feet above the river. At Burksville, in Kentucky, it is 125 feet. At Gainsboro Landing, in Jackson county, it is 332 feet. This shows a gradual dip eastward from Gainsboro and southward from Burksville.

Many of the high hills in the county are capped by a

soft, fine grained, yellowish sandstone, light and porous This sandstone is 25 feet thick at Burksville, in Kentucky. In Clay and Jackson counties it is rarely over six or eight feet thick. At places the Keokuk shales replace the entire Siliceous group, notably so at many places on Obey's river and its tributaries, Eagle creek and Ashburn's creek.

Soils.—The soils of this county will average better than those of Overton or Putnam, though the surface is so cut up by gorges and stream beds that a large amount of land can never be subjected to profitable tillage. Nevertheless, these rolling lands could be converted into pastures. In the eastern part of the county there are high hills which are the sloping ends of the spurs from the Cumberland table-lands.

The surface is broken, caves and sink-holes are common; and the soil is rich, lying on a strong clay subsoil. The hills and hollows, except where the land has been cleared, are covered by forests of large trees, among which walnut, beech, poplar, buckeye, linden and several kinds af oaks are common. Further west the Lithostrotion limestone continues to underlie the surface, often cropping out on the hillsides; the country is rolling, and the soil is generally a rich dark brown loam, with a red clay subsoil. North of Obey's river the red clay and limestone prevail to the valley of the Cumberland, and even west of the latter there are extensive areas of red rolling lands reaching nearly to the western boundary of the county. All of this red land is naturally rich, and with good management its fertility will never be impaired. Even when worn out by slovenly farming and constant cropping, its recuperative power is wonderful. Hickory, beech, sugar maple and dogwood are common on the hillsides and in the hollows or basins, and oak and chestnut on the hills and ridges. These red lands occupy a larger proportion of the area of the county than any other one class. Though not held in such high esteem as the river and creek bottoms, they possess many advantages over them. In other parts of this highland plain, particularly in the south and west, sandstones prevail, and the lands are less fertile. The red clay gives place to a yellowish subsoil, which is so hungry that the effect of manuring is scarcely perceptible after the first or second season. In some places on the hills are extensive beds of siliceous chert, known locally as "bastard flint." These gravelly soils are always leachy. Most of the timber consists of small posts oaks and black oaks. But even in these parts red clay and limestone, affording good lands, are found in spots. Small hickories are the prevailing timber in such places, and they are generally called "hickory barrens." In the north-west part of the county, on the head waters of a creek which flows north-west into Barren river the surface is more generally level, and there are some fine lands. In the valleys, the prevailing rocks are limestones of a different kind from that which appears on the surface of the highlands. They belong, geologically, to the Nashville Group of the Lower Silurian, consequently the soils are like those in the Central Basin. This limestone underlies all the valleys and outcrops on the sides of the hills about half way up on each side. It is highly fossiliferous, and by disintregation is continually adding to the fertility of the soil. In the valleys of the creeks, and also to some extent in the larger valleys, the soils have been modified by drift which comes down from the surrounding hills, so that they contain a larger proportion of sand than the same kind of soil otherwise situated. This sand mixed with the calcareous and argillaceous materials furnished by rocks, makes a very mellow, friable loam. Most of the creeks bring down also large quantities of chert, which gives a gravelly character to the soil where it is deposited. This gravel, however, rarely reaches out into the larger valleys in sufficient quantity to impair their quality. In the beds of all the creeks this chert is found in immense quantities. All along Cumberland and Obey's rivers there are alluvial bottoms of considerable extent.

These are naturally the richest lands in the county. The fluviatile deposits brought down by the river renew every year the waste of the soil, and some of these soils have for more than a half a century continued to produce crops of corn every year with no manure, and without any decrease in the amount produced. But there are some disadvantages to counterbalance these good gifts. Fences are often carried away by high water. Not unfrequently, when the fields are ready to be planted, a sudden freshet in a few hours obliterates the work of many days, and in some cases growing crops are destroyed by an unseasonable overflow.

Valleys.—In this connection, a particular description of some of the principal valleys may not be out of place. The largest and most important is that of Cumberland river. The part included in Clay county is fifteen miles long, with an average breadth of a little more than one mile. Crossing the State line a little east of north from Celina, it extends obliquely across the county in a direction rather more south than west. The numerous smaller valleys opening into it give to the encarpments on either hand a serrated character. The river meanders through the valley, often crossing from side to side, and many towering cliffs rise perpendicularly from the water's edge to the height of several hundred feet. In passing up or down the valley by land, it is necessary either to cross the river many times or to pass over the bluffs by rugged, toilsome roads. Obey's river valley is, in its general character, similar to that of the Cumberland, except that it is smaller. Reckoning from a few miles above the mouth of Wolf river, where it properly begins, it meanders first west south-west, and then a little north of west, to the center of the county, where it opens into the Cumberland Valley at Celina. Following its serpentine course, the distance is perhaps thirty miles or more, but in a direct line, not exceeding twenty. It has an average breadth, between the bases of the hills, of one-half to

three quarters of a mile.   Mill creek has a fine valley coming in on the cast side below Celina.   It is eight miles
long, and average nearly half a mile in width.   Iron's
creek valley, having about the same dimensions, comes into Obey's from the south, in the eastern part of the county.
The line of the proposed Southwestern Railroad passes
through this valley.   Kettle creek valley comes into
Cumberland from the north-west near the State line. about
three miles of the lower end being in this county.   It has
an avereage breadth of half a mile.   There are a number
of less important valleys, all of which contain good farms.
Of these, Mitchell's creek, Proctor's creek, and Brimstone
are the largest.

JACKSON COUNTY.

This county, lying below Clay county, on the Cumberland river, is in every respect geologically and topograph-
ically like Clay county, with few exceptions.   The hills are
more rounded, and the erosion has been deeper.   The black
shale appears higher on the hills, and the Cincinnati rocks
have a wider expanse.   Much more of it is in cultivation,
and the streams are more numerous.

There are more outcropping rocks, in some respect due to
bad tillage.   In consequence of the large number of streams
there is less plateau country, though a considerable scope of
this is to be found in the southeastern corner, adjoining Putnam county, also a few square miles in the northwest corner.
The remainder of the county is grooved deeply by streams,
some of them having extensive and valuable tracts of bottom lands.   The sloping hills furnish a fertile soil, which,
with proper treatment, might be made very profitable.
Seeded to blue grass these hills could be made sources of
countless wealth in the rearing of cattle and sheep, but in-
judicious cultivation is fast denuding them of the surface
soil, and leaving them sterile, presenting an aspect barren
in the extreme.

In coming from the direction of Cookville, over the plateau region, to the head of Flynn's creek, we find a flat, well wooded country, moderately fertile, soils siliceous and clayey ; identical in character with the clay lands of Putnam and Overton counties, and, indeed, a continuation of them· Descending the valley of Flynn's creek an excellent opportunity presented itself for making a section of the formations. Beginning at the top we have :

1. Siliceous rocks, chert and clay.................. 25   feet.
2. Keokuk shales.. ....·.............. 37½ feet.
3. Black shale............................. 18   feet.
4. Blue and gray, violet and buff colored shales,
      calcareous........................ 10   feet·
5. Layers of soft sandstone, interstratified with
      black earthy material........................ 10   feet.
6. Shales, buff colored limestone, weathering
      easily.. 18   feet.
7. Hard   bluish limestone, abundantly filled
      with orthis lynx and remains of cryptoceras
      undutum... 20   feet.
   In all to bed of stream............................ 138½ ft.

The limestones here abound in caverns. The rocks are fissured in an unusual degree. A mile below the head of the stream the extraordinary disturbance in the strata,which I have already mentioned, begins. There are numerous plaitings and foldings forming sharp curves and wrinkles on the left of the stream. Lower down on the right the dip in the strata is as great as sixty degrees, though the plaitings disappear,'and this extends for half a mile. At the end of this disturbance there is a fault by which the strata are dropped 200 feet, but retaining their horizontality. The line of fault is distinct in Cub hollow, near Antioch Church, and can be traced all the way up the side of the hill It crosses the hollow at right angles, the black shale lying side by side with the Nashville rocks.

Another region of great disturbance is found on the opposite side of Cumberland river, on Wartrace creek. Here
on the lands of Dr. A. M. Ferguson and L. H. McCarver,
are upheavals, extending over forty acres. Great hog back
ridges occur and appear to drip towards a common centre.
·Near this place, and lying at the foot of a long rocky spur,
which points down towards Wartrace creek is a singular
outcropping of quartzose sandstone covering an extent of
five or six acres. The line separating this sandstone from
the Nashville rocks can be traced with great distinctness on
two sides. It resembles in every particular the coarser
sandstones of the Cumberland Mountain, though, perhaps,
not so much stained with the oxide of iron. It is of a
slightly grey color, and would doubtless make a good sand
for the manufacture of glass.

Leaving out the Cumberland, Roaring river is the largest
stream in the county. It flows in a westerly direction
through the eastern half of the county. It is bordered by
fine rich bottoms from three hundred to eight hundred
yards wide. These bottoms are overlooked by abrupt hills
and ridges, rising from three hundred to four hundred feet
above the stream bed. During a rainy season it becomes a
very impetuous stream, overflowing the low lands, and oftentimes covering extensive tracts of arable land with sand
and gravel. In these overflows the river frequently cuts
out new channels through the bottoms, the banks being very
unstable and yielding. The soil of these bottoms is of a
superior character, yielding abundant crops of corn, wheat,
and hay. The average yield of corn is said to be fifty bushels per acre. The soil on the slopes of the bounding ridges,
derived mainly from the disintegration of the Nashville
limestones is largely intermingled with the cherty material
that tumbles down from the overlying siliceous rocks. The
slopes are often rugged with projecting layers of limestones.
At a point six miles above its mouth I measured one of
these bounding hills and found it to be three hundred and

thirty-five feet high, the Nashville limestone extending upward for two hundred and sixty feet, after which there were twenty-eight feet of black shale, capped by the siliceous group.

Blackburn's Fork, the largest tributary of Roaring river, except Spring creek, rises in Putnam county, and flows north, emptying into Roaring river, six miles above its mouth. The hills press close down to the water's edge on this stream, but it furnishes some valuable water privileges. The rocks dip locally in many places. At the old salt well, some two miles from the mouth of the stream, there is a gentle synclinal. This well was bored at first to the depth of one hundred and fifty feet in the Nashville rocks, and oil is said to have covered the drill. It was then sunk 199 feet and six inches, and strong salt water obtained. The salt manufactured from this water was not good, being tainted with the smell of petroleum. Six miles above the salt well are some beautiful falls, much resembling Waterloo Falls heretofore described. The water falls within a distance of thirty yards, sixty-five feet. The black shale here is twenty-six feet thick. Under the sheltering rocks of the precipice below the falls are found incrustations of alum and copperas in great abundance. These falls are on the lands of M. W. Cummins, and if utilized would make water power sufficent to propel any desirable quantity of machinery. A fall of seventy-five feet could be secured for a turbine wheel by working down the bluff on the northern side. Above the falls the water is very swift, falling twenty-four feet within the distance of a mile.

A section taken at the falls shows :

Surface	3 feet.
Siliceous limestone..	66 feet.
Black shale..	26 feet.
Nashville rocks..	35 feet.
Total height..	147 feet.

Bowerman's branch rises in the highlands, flows south, and empties into Blackburn's Fork, a short distance below the falls. This stream has some good water power though small. The streams on the south side of the river, beginning on the east that empty into the Cumberland are Roaring river, Morrison's creek, Doe creek, Flynn's creek, and Martin's creek. Those entering the Cumberland on the north, beginning on the east, are Jenning's, Hensley's, Cub, Bullard's, Indian, Wartrace and Salt Lick. They all have a general direction southeast, and have, with one exception, the same characteristics; that is to say they rise in the highlands, and scoop out deep valleys in the limestone with lofty ridges between. The bottoms on the streams constitute the best farming land in the county. Corn, oats, wheat, tobacco, clover, and the grasses constitute the principal crops. A comparatively small amount of ridge land is cultivated. These ridge lands are well adapted to fruit. The severe winter of 1876–77 destroyed all the peaches, and nearly all the peach trees in the low lands, but upon the ridges the trees literally broke down under the heavy weight of fruit. A fine quality of tobacco is also made upon the ridges, but not so many pounds are raised to the acre. Wheat also has a stiffer straw, owing to the predominance of siliceous matter in the soil. The following will show the comparative production and prices of the two classes of soil:

	Valley Land.	Ridge Land.
Wheat	10 bushels	6 bushels.
Corn	40 bushels	20 bushels.
Tobacco	900 pounds	500 pounds.
Hay	1½ ton	¾ ton
Potatoes, Irish	75 bushels	100 bushels.
Oats	30 bushels	20 bushels.
Price of lands	$30 to $40	$2 to $6.

Jennings creek presents some singular features that deserves mention. The valley will average three-quarters of

a mile, from the foot of the hills. The stream bed is filled
with sand and debris to such an extent as to absorb the
whole volume of water in a few days after the heaviest
rains. One may ride for half a day up the stream bed in
summer without seeing a single pool of water. These sands
shift from side to side, making torrid desolations totally unfit for agricultural purposes, amounting to nearly one-third
of the whole valley. The remainder of the valley lands are
considered the most valuable in the county, and sell for 30
or 40 dollars per acre. The fertility of the bottom lands
are preserved by the washings from the hillsides.

FENTRESS COUNTY.

Fentress county resembles Overton in its main features,
though a large proportion of it belongs to the Cumberland
Table-land. The whole of the eastern half belongs to the
carboniferous formation and forms and extensive plateau
with open pine woods, where the wild grasses grow in prod-
igal profusion during the summer months. The streams
flow in " rocky gulfs," varying in size and depth according
to the size of the stream. The soil in other portions of the
county is sandstone, loose, porous, leechy and unproductive,
chiefly valuable for fruit, timber and highway pasturage.
The western half is much serrated by bold spurs and intervening valleys or "coves," and generally marked by a
broken line of sandstone or conglomerate cliffs. From the
base of these, there is a steep declivity cut in many places
by deep ravines, and mostly covered with loose masses of
rock. The terrace, which is a characteristic feature of the
western aspect of the mountain in White and Van Buren,
is not so distinctly marked in Fentress, but the spurs projecting between the valleys of the principal creeks and
rivers occupy much of the county's area. These spurs
have the same elevation as the terrace, which is about half
the height of the Table-land. In places, however, there
are bold rocky hills rising high above other parts of the

range, and sometimes reaching an elevation equal to the
Table-land. The tables of these spurs are, in places, several miles wide, and there are some good farming lands on
them. This is especially ,true in the neighborhood of the
limestone knobs spoken of above. The lower slope, both of
the main mountain and the spurs, are often steep and
broken, but not generally so rocky as above. They are
generally covered with heavy forests of valuable timber,
but cleared fields are occasionally met with. It is a difficult
matter to trace the line which marks the base of the mountain. Unlike the escarpment above, there is no line of
bold bluffs—no natural boundary, but the smaller spurs run
out into hills and gradually melt away into the general
level of the Highland Rim. The valleys lie between the
projecting spurs, occupying, in the aggregate, about one-
fourth of the area of the county. There is considerable
variety in the surface and soil. In some places are extensive bottoms, while in others an undulating surface with a
red clay soil predominates. Taking the valleys altogether
as one natural division, we think that about two-thirds of
the area is clay upland, while the other third is divided
about equally between the coves and bottoms. Most important of these is the valley of Wolf river. Reckoning
from the place where the Three Forks unite to form
Wolf river, where it has a breadth of three miles, it extends north-westwardly, spreading out ten miles wide seven
miles lower down, and then grows narrower again. Its entire length in this county, is fifteen miles. The Three
Forks of Wolf is famous far and wide for the fertility of
its soil. Each of the three forks has a valley of its own.
That of Main Fork is about four miles long, and from a
quarter to a half mile wide. Middle Fork valley is about
the same size, while that of Rottin's Fork is somewhat
smaller. In all of them there are excellent lands. The
valley of East Fork is the largest in the county. The
head of it in the southwestern part of the county, is very

narrow, being nothing more than a "gulf" deep and rugged, and hemmed in by the almost precipitous mountain sides. But farther north, it gradually expands until it gains a width of six miles. Its length is about twenty-five miles. The river runs in a deep channel, while the surface of the valley is undulating, with good red clay soil. The lower slopes of the ridges, on either hand, are fertile, and in some places not too rugged to be cultivated. Indian creek, a tributary of East Fork, has a valley six miles long, by half a mile wide, similar in its general character to the larger valley of which it is an outlier. There are a number of minor valleys, lying between the various spurs and ridges, of which Dry creek valley is most important. It is three miles long, by an average width of five eighth of a mile.

4

CHAPTER III.

OIL SPRINGS ON THE UPPER CUMBERLAND.

HISTORY OF OIL DEVELOPMENT IN THAT REGION, WITH
RECORDS OF WELLS.

We have thus given a general outline of the principal
geological and topographical features of the counties on the
Upper Cumberland that is known to be oil producing.
We propose now to enter into a detailed account of the oil
indications, giving a record of the wells which have been
bored in this region and with what success.

Oil on Spring Creek.—The place where the most successful boring have been made for petroleum is on Spring
creek, a tributary of Roaring river. This stream rises in
Putnam county, and flows in a northwesterly direction,
forming for a considerable distance the disputed boundary
between Overton and Putnam counties. The wells were
bored one mile above the point where the public road leading from Cookville to Livingston crosses the stream, ten
miles northeast of Cookville, and about the same distance
south of Livingston. The distance from this point to
Butler's Landing, the nearest accessible point to Cumberland river, is nineteen miles, and the distance to Gainsboro
the county seat of Jackson county is twenty-five miles.

The following section taken in the region around the oil
wells at Spring Creek, by Dr. Safford, but since corrected to
correspond with the result of the borings, will serve to il-
lustrate the geological, lithological, and paleontological features of this section. The section given is really a combin-
ation of two sections, so as to include both the carboniferous, and the subcorboniferous formations.

SECTION AT SPRING CREEK, OVERTON COUNTY.

COAL MEASURES.

(3) *Sandstone*, on a high point south of the road. Thickness ?

(2) *Shales*, a heavy bed, with clay iron stones. This, with the sandstones, was estimated to be..................................130 feet.

(1) *Sandstone*, upper part thin-bedded or shaly...120 feet.
In all................................250 feet?

MOUNTAIN LIMESTONE.

10) *Blue Limestone*,........................ 4 feet.

(9) *Variegated Shale*, brown, gray, and green...12 feet.

(8) *Shale and Marl*, mostly gray, with some brown and green at top; at intervals some thin layers harder than others................ 40 feet.

(7) *Argillaceous Limestone*, dull bluish gray, breaking with conchoidal fracture; has cavities containing *dolomite*,............ 27 feet.

(6) *Blue Limestone*, fossiliferous,............ 22 feet.

(5) *Argillaceous Limestone*, resembling 7, above, but more compact, and somewhat fossiliferous.......................... 20 feet.

(4) Blue Limestones,........................ 85 feet.

(3) Shales, ... 6 feet.

(2) SANDSTONE, fine-grained, more or less flaggy,.. 48 feet.

(1) Blue Limestone, fossiliferous,..............168 feet.

Entire thickness,.......................432 feet.

LITHOSTROTION BED.
Upper siliceous

(2) *Cherty Limestone*, limestone not seen ; chert abundant on the surface............... 128 feet.

(1) *Limestone*, impure, of water-lime aspect, lower part containing sparry blue layers ; contains *Lithostrotion Canadense*.......... 75 feet.

In all................................203 feet.

PROTEAN BED.
Lower Siliceous

Sandstone, fine-grained, seen at a number of points in Overton and Putnam,............. 8 feet.
Limestone, blue, fetid, rather coarse, fossiliferous and crinoidal, seen...................... 45 feet.

Rocks penetrated by the boring of the Jackson Well; many layers chert,...................216 feet.

In all,...................................269 feet.

BLACK SHALE, resting on the Nashville Formation... 35 feet.

The wells are sunk near the eastern margin of Spring
creek, in a low flat place, inclined to be swampy. A wide
expanse of rolling lands spreads out eastwardly for about
three miles. On the south a bold spur comes up within a
few hundred yards of the wells with considerable gorges on
each side. On the northeast also, and on the west, considerable elevations are seen, the sides of which show thick
ledges of the St. Louis limestone. Lower down the limestones become very flinty and hard, and filled with horizontal crevices. The elevated points around the wells are from
fifty to one hundred feet high. The entire valley is about
three and a half miles long, by three quarters of a mile
wide, embracing the valley of Spring creek, and Hurricane, its tributary, which empties near the wells. The following cuts will show the geographical features about the
wells and the geological section.

SECTION OF THE HOOSIER WELL NO. 1, SPRING CREEK.

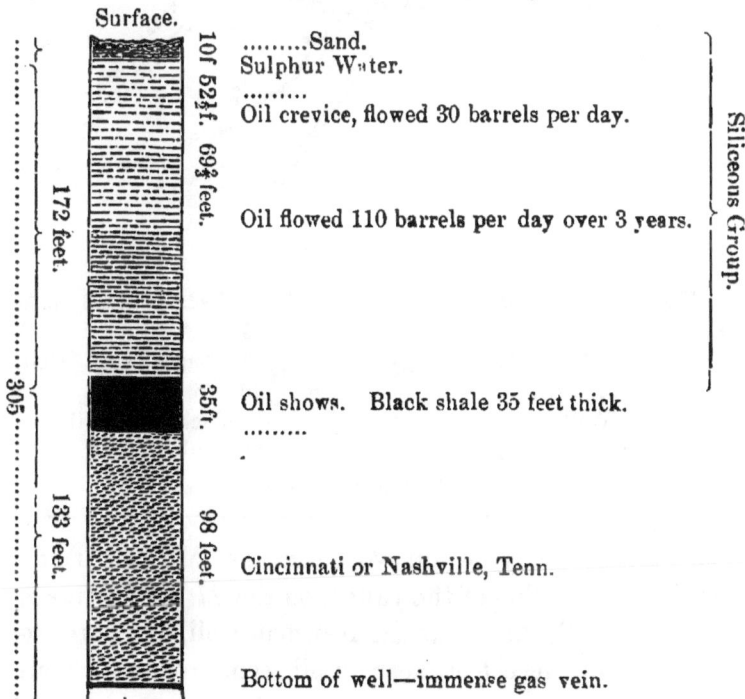

Surface.

.........Sand.
Sulphur Water.
....
Oil crevice, flowed 30 barrels per day.

Oil flowed 110 barrels per day over 3 years.

Oil shows. Black shale 35 feet thick.
....

Cincinnati or Nashville, Tenn.

Bottom of well—immense gas vein.

10f 52¾f. 69⅜ feet.

35ft.

98 feet.

172 feet.

133 feet.

305

Siliceous Group.

MAP OF REGION AROUND SPRING CREEK.

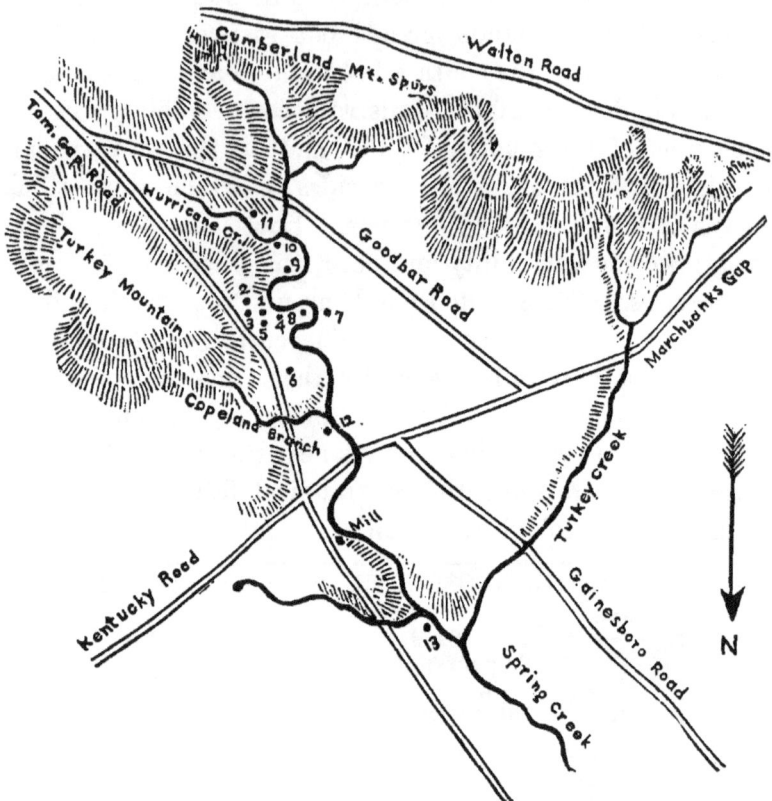

NAMES OF WELLS.

No. 1. Hoosier Well No. 1.
   2.   "   "   2.
   3.   "   "   3.
   4. Newman's Well.
   5. Douglas Well
   6. Pedrick Well, No. 1.

No. 7. Pedrick Well No. 2·
   8.   "   "   3.
   9. Hequemberg   Well.
  10. Roche Well.
  11. Jackson's Well.
  12. Netherland P. O.

13. H. L. Taylor's Well.

The first company began work on the McNeal farm, which forms a portion of the valley, on the 27th September, 1865. This well, known as the Newman well, was bored to the depth of nineteen feet, near an oil seep, from which oil had been exuding from time immemorial. Oil in a small quantity was struck at this distance, and the work abandoned until April, 1866, when the operations were resumed, under the superintendence of Col. Chas. H. Irvin. The well was deepened to twenty-six feet, and 2,600 barrels of oil pumped out. Sulphur water accumulated, and the well was abandoned until November, 1866, when it was drilled to the further depth of 51 feet 7 inches. At this depth the oil flowed out in great quantities, coming out in a terrible rush, and rising in a column thirty feet high. Thousands of barrels were lost, and it is said to have filled up a natural basin near the well, so as to be deep enough to swim a horse. The well continued to flow for three months. Twelve months afterwards the well was pumped, and six hundred barrels taken out. It was abandoned for a year, and then sunk to the depth of one hundred and twenty-six feet, when gas was struck, and the well permanently abandoned.

The next well was drilled on the Buck farm, lying on Hurricane creek, about three-fourths of a mile from the last. It was known as the Jackson well. This was bored 530 feet deep.

A section of the well showed:

Alluvial soil	7 to 9 feet.
Flinty limestone	165 feet.
Black shale	35 feet.
Cincinnati rocks	293 feet.
Total	530 feet.

A good show of oil was obtained at the depth of one hundred feet, but the rush of water was so great that no pumping was attempted.

The next was the Douglass well. This was bored sixty feet east of the Newman well. At twenty-two feet it yielded forty barrels of oil and was then abandoned.

The next was the Hoosier well, No. 1, on the Douglass farm. Work was begun on this well in the spring of 1866.

On the 13th of November, a strike of thirty barrels per day of forty-two specific gravity, was made. Depth fifty-two and a half feet. This supply continued for four weeks when the big strike, heretofore mentioned, was made in the Newman well which shut off the supply. Pumping was continued for four weeks, but no additional oil obtained. It was subsequently bored to the depth of sixty-nine feet and 8 inches and 110 barrels per day taken out. This supply continued for two years and three months, whenever pumped, but the difficulty of transportation made the pumping very irregular. Strong salt water came up with the oil. The product of this well was very light, sixty per cent being illuminating oil. Afterward the well was put down three hundred and five feet, striking the black shale at one hundred and seventy-two feet (the surface of this well was lower than the Jackson well). Gas was struck and the well abandoned. After two or three months the gas quieted down, but again came up when an attempt was made to deepen the well.

Hoosier well, No. 2, 110 feet east of No. 1, was also bored on the Douglass farm. At the depth of 55½ feet oil was struck, and twenty-five barrels a day were taken out. This well remained constant until work was abandoned at the place in 1871.

Hoosier well, No. 3, also on the Douglass farm, was sunk 350 feet north of No. 2, and 330 feet east of the Newman well. At 53½ feet oil was pumped out at the rate of 160 barrels per day. It continued to yield oil at this rate until it was deluged by the breaking in of sulphur water, which ruined it.

A well was bored on the Morton farm, the second above the Douglass farm. No oil was obtained at the depth to which it was bored, 72 feet.

The Hequamberg well, on the same farm, yielded no oil at the depth of 40 feet.

Pedreck well, No. 1, was sunk on the McNeal farm one hundred yards west of Hoosier well, No. 1. There was a good show of oil at fifty-four feet. It was carried to the depth of 126 feet. Torpedoes were then introduced and exploded but with no effect.

Pedreck well, No. 2, on McCullough's farm, two hundred yards west of No. 1, showed signs of oil at fifty-four feet. Torpedoes used with no resulting benefit. This well was bored on the west side of Spring creek.

Pedreck well, No. 3, on the McNeal farm, one-fourth of a mile north of the Newman well, was sunk 130 feet but no oil obtained.

All these wells were bored within an area of 150 acres, and the productive wells could have been enclosed in an area of four acres.

Two facts stand out with distinctness in the records of these wells :

1. The area of oil was very confined.

2. The oil was all obtained above the black shale, and was stored away in the crevices of the siliceous limestones.

In many portions of the oil region of Tennessee this siliceous limestone is filled with cavities, varying in size from a few inches in diameter to a foot or more. It is quite likely that the rocks about Spring creek, under ground, are perforated in a similar manner.

SHIPMENTS FROM SPRING CREEK 1867–9.

Very little oil was shipped from the Newman well, though it poured forth in such volume as to astonish all beholders. It is estimated that from 12,000 to 15,000 barrels run out of this well and was lost.

Dec., 1867.—From the Hoosier wells, Nos. 1, 2, and 3 there were saved:

Crude oil shipped.................................... 365 barrels.
Filled and not shipped............................ 135 "
   Jan., 1868, from second strike:
There were 231 runs of 14 barrels each.........3284 "
Other shipments by railroad and river..........1433 "
Twenty-four runs from large shute, of 50 bar-
   rels each, refined.............................1200 "
   July, 1869.—From well, No. 2, Hoosier..... 308 "
   From well No. 3 .... 138 "

Total product saved at Spring creek......6813 barrels.

At the time of my visit, in April, 1877, the oil was bubbling up and forming a scum upon the old Newman well. A company had begun to bore in the hard siliceous limestone, a mile lower down Spring creek. This company penetrated to the depth of about sixty feet, but from some reason, which I am unable to give, suspended operations· For much of the information pertaining to the operations on Spring creek, I am indebted to Geo. Satterfield, who had charge of the business for a considerable period.

Oil on West and East Forks and Obey's River.—Passing now in a northerly direction, through Livingston and beyond some ten miles, on the waters of West Fork, we enter upon an oil region in every way promising of profitable results. The West Fork hews its way down through the St. Louis limestones, Waverly sandstones and Keokuk shales. The bluffs rise, for the most part abruptly from the water's edge, with occasionally narrow strips of bottom land. The stream is one of great rapidity.

Koger's farm, upon which the oil indications are most numerous, is within four miles of the mouth of West Fork. The bluffs on the river show a succession of wrinkles, and the limestones are soft and much eroded by

atmospherical and plavial agencies. At places they are
vesicular and cavernous. The hills are oftentimes sloping
and covered by a coating of unctious clay. The following
section was taken on the east side of the stream, just above
where the oil indications are most numerous.

SECTION AT KOGER'S FARM ON WEST FORK.

On high points above, back from the stream, sand-
    stone, fine-grained and buff colored................20 feet.
Buff colored sandy limestone........................... 6 feet.
Gray crystalline limestone.............................79 feet.
Calcareous shales, blue and gray......................37 feet.
Thin beds of bluish and buff colored limestones, in-
    clined to be shaly—some few sandy layers......113 feet.
Sand and alluvial...................................... 6 feet.

 This last is a small bottom about 115 feet wide running
up and down the stream. The sands of this bottom are
thoroughly saturated with oil. By sinking a hole any
where in this small strip, or in the bed of the stream, petroleum rises to the surface in brownish or bluish disks,
which float away on the surface of the stream. The bluish
disks are beautiful in their irridescence, displaying in the
light all the colors of the rainbow. At this place many
gallons may be collected in a day. The oil-saturated sand
extends for the distance of one hundred yards or more. On
the east side of the stream, after ascending the bluff, there
is a level plateau extending two miles or more eastward in
the direction of East Fork. A considerable ridge then rises
up the water-shed, between East and West Fork, succeeded
by another plateau bordering East Fork. The ridge spoken
of is made up of mountain limestone, capped by the lower
sandstones of the coal measures. To the westward a like
plateau, though higher, is covered with a yellowish porous
sandstone, forming a thin unproductive soil. This continues westward comparatively level for six miles, when it
gradually rises into the towering heights of Pilot Mountain,

which stands a prominent and striking landmark. This mountain pertains to the coal measures, though separated from the Cumberland table-land by an intervale of eighteen miles.

At a point a little above the Koger farm, where the Livingston and Jamestown road crosses West Fork, oil occasionally oozes out from the bank of the stream, and just below are several places where salt water comes from the bluffs, leaving a salty incrustation upon the face of the rocks.

Below the Koger farm half a mile a saw-mill stands upon a short tributary stream, a few yards from West Fork. In blasting out rock at this place for the construction of a dam, oil was found in the pores and crevices of the rocks. Still lower down the stream and one mile above its union with East Fork, the mass of slimy mud on the side of the stream is thoroughly saturated with petroleum. By pushing a stick down in this mud, gas and petroleum ascend to the surface from the spongy mass. The very atmosphere at this place is redolent of petroleum, and one can scarcely touch the earth for a space of several hundred feet without having his olfactories offended by the odor of the oil. The formations here agree precisely with those at the Koger farm. The Waverley sandstone appears on the heights above in some greater thickness, and one layer of sandstone a foot thick is found interstratified with the Keokuk slates, which here take the place of the siliceous rocks.

Just below the junction of East and West Fork, about three quarters of a mile, a flow of oil comes out of the bed of Obey's river. This is on the lands of James Lacy. It is said that when the water is low the sand taken from this place is so saturated with oil as to make a bright flame.

At Goose Neck Bend, up East Fork, three miles above its junction with West Fork, there is an oil spring, and several more are reported to exist higher up the stream on the lands of Peggy Smith and Matilda Wright.

On both sides of the river at Lacy's farm the confining

hills rise to an elevation of over three hundred feet. On
the north is a solid and almost perpendicular bluff overlooking the river. On the south side there is a steep slope
up to the plateau lands which are here margined near the
top by sandstone, but this disappears a short distance from
the edge of the bluffs, and the plateau extends southward
to Pilot Mountain and westward to Eagle creek.

One half mile below Lacy's another oil spring breaks
out near the center of Obey's river. The oil has been seen
here for fifty years.

Three miles further down, near the mouth of Franklin
creek, oil is seen coming out of the bed of the river immediately under a bluff 300 feet high. The oil oozes out
beneath thirty feet of black shale. The rocks above the shale
at this point are filled with crinoidal stems in such abundance
as almost to make up the mass. The oil here is lighter than
that on West Fork and plays with its rainbow-hued disks
upon the surface of the water. Between the mouth of
Franklin creek and Lacy's farm the river cuts through the
black shale.

Franklin creek is in Fentress country, and runs west,
emptying into Obey's river. It is about three miles long.
One mile above its mouth there is a gravelly bar which is
steeped in petroleum. By riding a horse in the stream at
this place oil comes to the surface. The pebbles look as
though they had been immersed in petroleum. The oil is
above the black shale. The ascent from the mouth of the
creek to this place is fifty feet or more. There are several
long sags in the strata here, and greenish and bluish shales
are seen in the overlooking bluff. This oil seep is near the
ragged edge of the Cumberland mountain. Going eastward three miles a seam of coal four feet thick crops out.
It was worked for some time before the war. The coal is a
hard block, and of excellent quality. A quantity' of it taken
out seventeen years ago is still marketable, so well has it re-
sisted disintegration.

Near the mouth of Eagle creek there are many indications of oil. Seeps occur at intervals from its mouth for six or eight miles up the stream. Eagle creek, as will be seen by reference to the general description of Overton county, rises five miles north of Livingston, flows north and northeast and enters Obey's river, seven or eight miles below the mouth of West Fork. Its entire length in a straight line is about ten or twelve miles. It, like the other streams flowing into Obey's river, is walled in by bold bluffs of Keokuk shales and crinoidal limestones, which in this part of the oil region take the place of the siliceous beds. Near the mouth of Eagle creek the black shale is exposed and forms for a considerable distance the fissured bed of the stream.

Three miles above is an oil exudation on the right side. A well was bored near this place in the spring of 1866 to the depth of sixty feet, which yielded thirty barrels of oil. The oil was much heavier than that obtained on Spring creek, and found a lodgment in the porous limestones of the Nashville or Cincinnati group that lie below the black shale.

A second well was bored on the margin of the stream within a few feet of an oil seep, but no oil was found, although it was carried to a greater depth. The gas ejected the augers and drill from the well, and it was found impossible to continue the work.

Between this place and the mouth of the creek, and one mile above the mouth, two other wells were bored within a few inches of each other. These wells began on the black shale. Oil was found in both at the depth of fifty feet, but a larger supply at eighty feet. A considerable number of barrels, probably over a hundred, was obtained from this well, but the price of oil at the time got down so low that the transportation cost more than the oil was worth in the market, and the wells were abandoned. One of these wells was subsequently carried to the depth of 170 feet, but with no additional increase in the oil product.

Still nearer the mouth and below the black shale another well was sunk 300 feet deep. A small quantity of oil with a large flow of salt water came out of this well. So great was the volume of the latter that it killed all vegetation in its course to the river.

Just below the mouth of Eagle creek, on Robert's place, oil spurts out from the bed of Obey's river, and also at another point four miles below in the bed of a little branch on Jolly's farm. A well was sunk at this place and oil obtained.

Returning to Eagle creek. Nine miles above its mouth Whites creek enters on the left. Thirty years ago a well was bored on this stream in search of salt water, and after boring 70 feet a large volume of oil is said to have run out, much to the disgust of the salt hunters. Just how long it continued to flow, no one knows, but tradition assures us it continued for years.

Between the mouth of Eagle creek and the mouth of Ashburn's creek, eight miles below, there is an extensive plateau, four or five miles wide and ten long. The forests are dense, the soil of medium fertility, and the surface covered with fossils pertaining to the Lithostrotion bed of the sub-carboniferous. Wagon loads of the *Lithostrotion Canadense* may be gathered on this plateau.

Near the mouth of Ashburn's creek, in Clay county, the rocks in the bed of the creek have a decided dip towards the north. They are much fissured, and oil is often made to rise to the surface by stirring the accumulated mud which fills the crevices.

Numerous seeps occur in the margin of the stream for four or five miles above its mouth. The black shale forms the bed of the stream a short distance above its mouth, and appears to be saturated in an unusual degree with petroleum. Sulphur and chalybeate springs abound.

Salt water with some oil was obtained up a hollow a half mile below the mouth of Ashburn's creek, and one-fourth of a mile from Obey's river. The well was bored 178 feet deep, and about 10,000 bushels of salt were manufactured in the year 1867. Sixty gallons of water made a bushel of salt. The cost of making salt was 7½ cents per bushel, but bacon was then selling at 25 cents per pound, labor $1.50 per day. It could now be made at 3 cents per bushel.

Near the salt well, up a branching hollow, is an oil seep. The calcareous shales which here have a great thickness, are saturated with petroleum. Oil has been found oozing out at several places in the vicinity.

Wolf river enters the Obey's a few miles below Ashburn's creek, but on the opposite side. Nearly opposite the mouth of Wolf river, a short distance below, the strata exhibit a beautiful synclinal, the length of which is probably a fourth of a mile. The rocks are grooved with a singular and pleasing regularity, making the bluff appear like the rock mouldings of an immense superstructure. Nearly opposite this synclinal Parsley's creek comes in from the north. Upon this is an oil spring. On Wolf river, a mile above its mouth, a well was bored 60 years ago for salt water. Tradition tells of a greenish, oily fluid, highly odoriferous, running down the river, which was set on fire, producing a terrible conflagation.

Sulphur creek comes in also from the north, three miles below the mouth of Wolf. Several oil springs occur on this stream, but within the State of Kentucky. A well was dug near the mouth of Sulphur forty years ago by Mr. Trousdale, and a large flow of oil obtained.

At the mouth of Poor's branch, which enters Obey's river ten and a half miles above Celina, oil in low water is seen upon the surface. This comes from the Nashville rocks, the black shale appearing thirty-five feet above in

the bluff. Fossils abound in the limestone at this place.
Indeed, some of the layers appear to be little else than a bed
of fossils solidified. They are chiefly the *orthis lynx* and
orthis sinuata.

A well was dug near this place for domestic purposes, but
the water was so impregnated with petroleum as to be unfit
for use.

On Mill creek, below Celina, a well was bored in 1868 on
the farm of L. B. Butler. A considerable amount of oil
was obtained at this point and shipped, but I was not able
to ascertain the precise number of barrels. The oil from
this well came from the Nashville or Cincinnati rocks.

I have mentioned the oil spring on Franklin creek in
Fentress county. Others are found on the western edge of
that county in various places. There is a group of such
springs near the mouth of Poplar Grove creek and several
on East Fork, besides others reported, which are enough to
confirm the existence of petroleum in this county.

In Clay county, on Brimstone creek, which is near the
line which separates this county from Jackson, there are
several places where the oil exudes from the earth.

In Jackson county, at Allen's mill on Blackburns Fork
of Roaring river, oil is said to have been obtained in sinking a well for salt.

On Mill creek, in Clay county, petroleum oozes out from
the Nashville rocks.

On Wartrace creek, in Jackson county, also, in boring for
salt, solidified petroleum was met with in the cavities of the
Nashville limestones.

Just above the mouth of Jenning's creek, in the same
county, a well was sunk on Buck's branch to the depth of
forty-two feet, and some oil is said to have been obtained,
though how much it was impossible to ascertain.

I could hear of no oil indications in Smith county nor in
DeKalb.

I cannot close this chapter without returning my thanks

to the citizens of Overton county for the deep interest they manifested in my investigations, and especially to J. S. Roberts, Esq., and J. W. Wright, who did not hesitate to leave their respective vocations, and guide me to most of the oil indications in the county. My obligations are also due to John McMillan, Esq., of Celina, for similar courtesies.

5

CHAPTER IV

TROUSDALE, MACON AND SUMNER COUNTIES.

GEOLOGICAL PHENOMENA—MILK SICKNESS—KNOBS—
GLASS SAND—MILLSTONE GRIT—PASTURE LANDS—
OIL ON TRAMMEL'S CREEK.

Trousdale county is thought to contain deposits of petroleum. On Lick creek, a tributary of Dixon's creek, there are some interesting geological phenomena. The strata are exceedingly wavy, forming gentle synclinals from a few hundred yards in length to a quarter of a mile. Two miles above the mouth of Lick creek a soft, fine-grained sandstone presents itself, brownish in color, very porous, and about twenty feet thick. It is easily broken and will absorb water like a sponge, the water penetrating every part of it in a short time. When first dipped in water a bluish film floats away from it much resembling petroleum. A well was bored a mile from this point four hundred feet deep, and water spouted out, rising above the tops of the trees, which ran into a mill-pond and destroyed all the fish. Deeper boring was attempted, but the auger became fastened in the crevice of a rock and the well was abandoned.

In the bed of the creek near where the sandstone makes its appearance on Jo. DeBow's farm, the odor of petroleum when the water is low, is said to be very distinct. All this portion of Trousdale county, that is to say the northeastern corner, is very rough and cut up by deep hollows, and abounding in sharp-crested ridges from three hundred to four hundred feet high. The soil is very fertile, even on the slopes and crests of these highlands. The black shale occurs near the top, and the black chert lies thick upon the summits, so as to interfere to some extent with proper cultivation. Many of these ridges resemble great railroad embankments, being very symmetrical and narrow, oftentimes just wide enough at top for a good road.

On the ridge dividing Pumpkin creek from Lick creek a coarse quartzose sandstone is found in ledges about two hundred feet below the top of the ridge. But even where this sandstone prevails the soil is fertile, as is shown by the luxuriant growth of pawpaw bushes and tangled masses of creeping vines. Beech, sugar tree, chestnut, chestnut oak and poplar constitute the prevailing timber.

Northwest of Hartsville rises to the height of five hundred and forty feet above the Cumberland river, a series of bold knobs remarkable for their sandstone and mill-grit formation. These knobs are covered thickly with cane, through which it is almost impossible to walk. Forming the southern escarpment of one of these knobs is a great unstratified ledge of coarse-grained white sandstone about fifty feet thick. It stands out like a solid wall. The sandstone is easily crushed, and has been tried by glass manufacturers and pronounced of superior quality.

The mill-stone grit lies just below the black shale in an adjacent knob. It pertains geologically to the Nashville series of rocks, and consists of a mass of silicified shells closely compacted. Where exposed the calcareous material has been leached out, leaving the mass cellular. The stratum is about seven or eight feet in thickness. The millstones made of this material are highly prized for grinding corn, the sharp points and edges cutting rather than crushing the grain, whereby its original sweetness is preserved. For some years the manufacture of these stones formed no unimportant branch of industry.

Baryta is found in great quantities and of excellent quality half a mile south of the bluff of sandstone.

Galena is said to exist in considerable quantities in the county. I saw some excellent specimens of this ore said to have been obtained in the county.

For pasturage these knobs would be unrivaled but for the presence of that inscrutable agent which produces milk sickness. Observations for a number of years here show that the largest number of cattle is attacked by it just after the first frost in autumn, and just before vegetation puts out in spring. From this circumstance it is supposed to be due to some plant that makes it appearance early in spring, and is hardy enough to resist the first frosts.

Nature teaches us a lesson under the sandstone ledge, as to how soils may be improved. Great blocks have broken off and rolled down in the valleys. Constant erosion is going on, the loosened sand commingling with the clayey soils in the valley. This has so ameliorated the soil on the slopes and in the valley that it produces the most bountiful crops. It is loose, easy to work, and its porosity prevents washes, while it enables the soil to preserve a due degree of moisture. Lime is also added from the limestones that underlie the sandstone, so that all the elements of fertility and amelioration are here presented.

With the exception of the knobby portions, there is no county in the State blessed with a more productive soil than Tronsdale. It is a blue-grass region, and its sloping hills and wide valleys should be carpeted with greensward, and covered with the finest breeds of sheep and cattle, attesting at once agricultural thrift and refinement of culture. It is a postive prostitution to sterilize these rich slopes by the culture of tobacco. The soil is not adapted to the growth of that weed. The quality produced is thin, papery, and almost destitute of nicotine, the active principle of tobacco. Nature has pointed out the tobacco regions of Middle Tennessee as clearly as she has defined the limits of the succeseful stock-grower. The Central Basin is essentially stock raising. No cotton or tobacco should ever be planted in it. To do so is like condemning a lady of winning address and refined culture to be the scullion of a kitchen. On the other hand, the clayey soils of the Highland Rim produce and very finest grades of tobacco, but they are not adapted to the grasses. or at least but poorly so, and even then to a few select varieties. To attempt to clothe our clay hills in blue-grass is like attempting to make a fine lady of the kitchen scullion. It will not last. Many a hill-side in Trousdale county has been ruined beyond redemption by bad tillage, and is now not worth the taxes paid on it. The bare ledges of limestone rocks reflecting the glare and radiating the heat of the sun, make painful pictures in the landscape, and mark the want of true agricultural wisdom. Smith and Jackson counties tell the tale of the same sad treatment. Wherever these slopes have been seeded to blue-grass they make a picture of unsurpassed beauty, and become positive sources of wealth without the ceaseless efforts of man. When once the rocks have been denuded of soil, as they have often been in these counties, fifty generations cannot replace it. Such hills remain everlasting monuments of the folly of man.

The true policy is to make these sunny slopes of perpetual value, by either leaving them in timber, or thinning out the trees and seeding to blue-grass, for which they are so well adapted. By this means the value of the lands in the county would be increased, the freshness and beauty of the landscape enhanced, the profits and pleasures of the farmers enlarged, the system of earth-butchery arrested, and the present generation would not be held responsible for a devastation more alarming and more wide-spread than that produced by all the fires and storms and earthquakes and wars that have occured since the settlement of the State. The evil is the more dangerous because it is silent and insidious, though perpetual. The changes for the worse are so gradual and gentle that men come to look on them as they do upon the creepings of old age; but unlike the latter, they may be arrested and arrested before the doom of poverty is entailed upon our children. The soil should never be so treated that each year makes it more difficult to live the next. This is not civilization, it is postive barbarism.

Some few indications of oil are found in Macon county. Near the Red Sulphur Springs it oozes out from the black shale formation. The county has more plateau land than Jackson. Indeed, nearly the whole of it belongs to the sub-carboniferous formation. Toward the north there is a wide area of level and gently undulating land. It inclines gently toward the north, sufficiently so to give a good flow to many of the head branches of Barren river. There are also inclinations to the east and south—numerous small streams flowing in these directions to the Cumberland river. On the west the highlands break off in steep declivities, which run down into deep valleys exposing the Nashville rocks. The greater portion of the county is underlaid by the black bituminous Devonian shales, and sulphur springs abound in many places. The soil is usually clayey and thin, but there are extensive areas of great fertility. The county is well wooded.

Sumner county, lying west of Macon, shows some signs of oil in a region of country lying upon Little Trammell creek. Several oil wells were bored on this stream, some of them in Kentucky, and oil in small quantities was found in all of them at the depth of 30 feet. The daily product from the only one that was pumped, did not exceed two barrels. The entire production from this well was about 200 barrels. This well was afterward sunk to the depth of 500 feet with the hope of an increased yield, but no other source of supply was found. Gas and salt water were found at 250 feet, the water coming up with great force, rising in a column fifty feet high. The gas and water filled an underground cavern, the augur dropping two feet just before the column of water began to flow from the well. The source of the supply was the black shale, the oil being struck near the bottom of that formation. The product was a very heavy lubricating oil, which was used by the Louisville and Nashville Railroad Company, and said to be very superior. It sold for nearly double the price of the lighter oils.

Some oil has been found in Davidson and Maury counties, but not enough to justify expensive exploration. A company is now boring on Brown's creek, near Nashville, in search of oil.

CHAPTER V.

OIL IN DICKSON COUNTY.

GEOLOGY—HISTORY OF OIL-WELLS—PRODUCTION, ETC.

It remains to give some account of the oil operations west
of Nashville, particularly in Dickson county. This county
is so far removed from the oil region, of which Overton
county is the centre, that I have not included it upon the
map accompanying this report. Dickson county belongs to
the same great natural division of the State that embraces
the larger portions of Overton, Putnam, Clay, Jackson,
White and Macon, that is to say, the Highland Rim. In
nearly every respect, geologically, the county resembles
Clay. It is a broad plain furrowed by numerous streams
which cut down through the siliceous group of the sub-
carboniferous rocks. There is this difference, however. In
Dickson county in many places the lower member of the
siliceous group presents itself as a pale yellowish porous
sandstone. Sometimes great bluffs of this rock are seen
along the water courses. Some slight traces also, of the
Niagara rocks of the upper Silurian, under the black shale,
are met with. The upper Silurian, as has been stated, is
wholly wanting in Overton, Putnam, Clay and Jackson
counties. In the western part of Sumner, and probably in
Trousdale county, the Meniscus gray limestone of the Niagara epoch of the upper Silurian, is present in considerable
volume, being estimated at about 120 feet in thickness, on
the Gallatin and Glasgow turnpike, where it ascends the
ridge and leaves the basin.

On Jones' creek, in Dickson county, about seven miles north-west of White Bluff, a station on the Northwestern railway, three wells have been sunk on the farm of Mr. G. W. Brown, just below the black shale formation. These wells were bored about the same time (1866–69) when the developments before described in this pamphlet, were making in Overton county. The first was bored to the depth of 187 feet, which only resulted in striking a fissure from which issued a copious flow of gas. It would have been bored deeper, but a part of the drill was lost, which it was found impossible to recover. Another well was then sunk, eleven inches from the one just mentioned, to the depth of 295 feet. At this depth some oil was obtained, which flowed immediately after striking it, at the rate of thirteen barrels in a half an hour. The pumping apparatus was then applied and a considerable quantity of oil obtained, when finally it ceased flowing, but an unintermitted flow of gas continued. The well was now abandoned and a third one sunk about fifty feet from the location of the two first. This was bored to the depth of 565 feet, and from it came the greater part of the oil obtained in this region. The fuel used in drilling the last mentioned well was the gas which issued from the second well sunk. No account was kept of the total amount of oil obtained, but from 200 to 300 barrels it is said were shipped to Nashville by railroad, where it was refined and sold. The gravity of the oil here is about 42°.

Concerning the work at present at this place, the following may be stated : During the early spring, when the leasing of southern territory engaged the attention of northern operators, some persons from New Castle, Pa., re-leased this territory from those who succeeded the old company, and put up a rig which was completed about the middle of June. Since that time work has gone on gradually until my visit July 1. The well had been sunk to the depth of 68 feet. At the depth of about 57 feet gas was struck, which con-

firms the facts relative to operations here in former years.
The superintendent is very sure of getting a paying well,
and holds that the rocks here are similar to those of the
Pennsylvania region.* The formation is altogether different,
though the lithological features of the valley may be somewhat similar. The surroundings of the wells deserve a no-
tice also. They are located at the mouth of a ravine which
makes a decided indentation in the highlands, the foot of
which is, barometric measurement, 230 feet below the general
level of the rim-land plateau. The wells are about 150
yards from the south bank of the creek, which winds its
way in a general north-east direction, emptying into Big
Harpeth. The creek at this place skirts the southern side
of a beautiful little valley which comprises an area of about
150 acres of very fertile land. It is hemmed in on all sides
by the highlands, which form a wall 200 feet high, being
broken apparently, only by the cove-like ravine at the foot
of which the wells are located, and at the two points from
which the stream makes its entrance and exit.

Some indications of oil have been found in Hickman and
Montgomery counties. In both counties some wells were
sunk about ten years ago, but without any profitable results.
In Hickman county the Meniscus limestone furnishes some
oil. At Montgomery's mill, on Piney river, a black looking petroleum has been oozing from this formation for nearly
half a century. It was first discovered in blasting out rock
for the foundation of the mill. It is collected in small
quantities for medicinal purposes.

At Centreville oil oozes out in a drain near the base of
the hill between that place and Duck river. It also appears at the base of a hill on the south-east of Centreville,
near the valley of Indian creek.

Leases.—Nearly all the lands in the oil territory are leased,
the lessee agreeing to pay from one-sixth to one-twelfth of

*Oil was found August 25, 1877, at a depth of 445 feet.

the gross production to the owner, and to begin operations within a specified time, varying from two to five years. Some few have refused to lease, but have signified their readiness to make a liberal arrangement with any one who is ready to begin work, and who will bring suitable apparatus for boring.

CHAPTER VI.

___

OTHER RESOURCES OF THE OIL REGION.

___

PRODUCTIVITY AS COMPARED WITH THE WHOLE STATE—
TOBACCO — WOOL—GRASSES—IRON ORES—COAL—LEAD
—WHETSTONE GRIT—MINERAL WATERS—WATER POW-
ERS—LUMBERING TRADE, PRICE OF LUMBER, LABOR—
LETTER FROM HON. J. D. GOODPASTURE.

The soils have already been mentioned. Their capacity
of production is very considerable. It is the general impression that all oil regions are barren and unproductive,
but in this particular the Tennessee oil region is an exception, as the following table will show :

	PER CAPITA PRODUCTION FOR STATE.	PER CAPITA PRODUCTION FOR OIL RE- GION.
Corn........................	32 bush.	37 bush.
Wheat........................	5 "	3 "
Oats........................	3	5
Tobacco........................	17 pounds	30 pounds
Wool........................	1 1-10 "	2 "
Honey........................	4-5 "	2 "
Sorghum........................	1 gal.	2 gals.
Animals sold for slaughter, value per capita of population...........	$13	$15
Live stock, value per capita of population........................	$44	$46
Sheep, number........................	$\frac{2}{3}$	$1\frac{1}{3}$
Swine........................	$1\frac{1}{2}$	$2\frac{1}{3}$
Improved land........................	5 acres.	6 acres.
Unimproved........................	10 acres.	14 acres.

The comparison is made between the whole State and the central portion of the oil region embracing four counties, Clay, Jackson, Overton and Putnam.

Several important facts are brought out in this table. It will be seen that in all the staple farm products, with the exception of wheat, this region, in proportion to population, far outstrips the average for the State. The small amount of wheat grown is due to want of the means of transportation. No effort is made by the farmers to raise more than is required for home consumption. It takes twenty cents per bushel to send wheat from the vicinty of Livingston to Cumberland river, and about the same to transport it to market at Nashville, so that when wheat is selling at the latter place for one dollar per bushel, the farmer near Livingston realizes only sixty cents, without estimating any charges for commissions or storage. On soils making an average yield of ten or twelve bushels per acre, sixty cents per bushel would not be a reasonable compensation for the work expended in making the crop.

The production of oats and corn is far above the average. This is fed to hogs, cattle and mules, which are driven out on foot to market.

Tobacco, as a money crop, presents a great many advantages for an inland county, because it brings more money, pound for pound, than any other farm product. This reduces the expense of hauling. Two hogsheads of tobacco weighing 1,600 pound each, will require two days to get them to the river from a distance of twenty-four miles at a cost of eight dollars. The charges from the river to the warehouse in Nashville, is about four dollars each, making the total cost for putting a hogshead in market eight dollars. But an average hogshead of tobacco will bring $125, so that the cost of transportation is only one-sixteenth of the whole. The proportion between the cost of getting a hogshead of tobacco to market and the value of the tobacco, is less than for any other farm product. Overton, Putnam, Clay, and

a portion of Jackson, have suitable clayey soils for the raising of this crop, and upon such soils it should be grown,
but never upon soils that will produce blue-grass, nor upon
soils easy to wash.

Another article that might be produced with great
profit is wool. The rolling surface of this region, the
the sheltering hills, the wild grasses that spring up sponta-
neously everywhere, the adaptability of the soil for the production of the domestic grasses, particularly blue-grass,
herds-grass, orchard-grass and clover, and above all, the
cheapness of the lands, point out this locality as one where
sheep-raising and wool-growing might be carried on with
great profit to the farmer and with benefit to the soil. The
number of sheep, under the operations of the much villified
"dog-law," constantly increased, and a few more years
under the benign protection of that law, would have seen
the rolling heights and sheltered valleys of this section of
our State covered with flocks whose "hoof is gold." The
proportion of the number of sheep to population was just
twice as much in 1870 as the average of the State. The
number of sheep then reported for this district was over
43,000. The number now is probably 55,000.

It will be seen that though there are six acres of improved land for each inhabitant, there are fourteen unim-
proved. The system of old and new field culture which is
always practiced in a thinly settled region, is carried on to
a great extent, particularly in Overton and Putnam counties, not so much in Jackson and Clay. New fields are
opened every year, and old fields are turned out to grow up
in bushes and briers. There is no system of rotation or
rest carried on. But few farmers pay any attention to ma-
nures, or make any efforts to keep their soils in good heart
by sowing clover or seeding to grass. Crop, crop, crop, is
taken off year after year, suggestive of Hood's "stitch,
stitch, stitch," until the poor victimized soil refuses longer
to respond to the labor of the farmer. It is then cast aside

as worthless. Yet there are no soils which show the effects of manure quicker, or upon which it will last longer than the clayey soils of Clay, Overton and Putnam. Hon. J. D. Goodpasture, in the year 1876, took the poorest field he had, which had, however, a good under-clay, broke it up deep, listed it and applied stable manure to the hill. He gathered fifty bushels of corn per acre from the field, exclusive of "nubbins." Judge W. W. Goodpasture made some experiments with hay a few years ago in Overton county, with excellent results. From four and a half acres sown in timothy and clover, he obtained at first cutting 31,000 pounds of hay. At the second cutting the same season, 15,000 pounds were obtained, making in all for one season 46,000 pounds, or 26 tons, a little over five tons per acre. About equal parts of timothy and clover were sown. The timothy and clover grew as high as a man's waist and was very thick on the ground. This crop was raised on land that had, as the phrase is, been "worn out," and was considered poor land, having a clay sub-soil—a light clay loam.

On about two acres of this land hogs had been pastured and fattened for two years. On the other portion of the four and a half acres stable manure was spread broadcast quite freely. On the part fertilized with hog manure, the yield was better than on the other part. The previous year this ground had produced a heavy crop of oats, the clover and timothy having been sown with the oats.

The usual average of corn for the clay uplands is about twenty-five bushels, wheat eight bushels, oats twenty bushels. About a hundred gallons of sorghum is the product of an acre; sweet potatoes yield well. One hundred and seventy-five bushels have been raised upon a single acre. Apples and grapes bear with great certainty. Peaches rarely fail on the uplands. Bees find abundant food in the poplar blossoms and in white clover. On the bottom lands and

in the coves the yield of all field crops is about twenty-five
per cent. greater than on the uplands.

Farm labor is cheap.   Fifty cents per day and one hundred dollars per year, with board, are the usual prices.

Iron Ores.—Throughout Overton county and a small part
of Putnam and Fentress counties nodules of red hematite
are picked up, called dyestone, which will yield about 65
per cent. of metallic iron.   It occurs in the siliceous group,
and is found associated with chert.   It is the only red hematite which I have met with west of the Cumberland Table-
land, except the deposit at Clifton.   The specimens which
I found are angular, as though they had been shivered by a
blow.   At some places a bushel of the fragments have been
picked up.   I found the largest quantity on the east side of
the eastern road to Cookville.   It was formerly in great
demand by housewives for dyeing cloth.   I do not think it
exists in any quantity, though a fine deposit is said to exist
at Ramsey's mill, on East Fork, one and a half miles west
of the line of the Pacific railroad.   On Aleck Verbal's
land, on West Fork, four miles west of Ramsey's mill, is
another reported deposit.

Brown hematite is found in several localities.   Southwest
of Livingston a considerable bed exists.   It was dug for
many years on Town creek, and used in a Catalan forge on
Roaring river near Crawford's mill.   Some masses were
found large enough to make several loads for a wagon.

Eleven miles north of Livingston, on James Sell's
place, on the headwaters of Ashburn's creek, brown hematite is found in some quantity, probably enough to justify
the erection of a forge.   The ore is honey-combed and generally free of flint.

Beds of stratified siliceous ore are met with on Puncheon
Camp of West Fork, on Martin French's place.

There are extensive beds of brown hematite around Pilot
Knob, in Putnam county; also in places near Cookville.

A large deposit is said to exist near the oil wells on Spring creek.

Coal.—Coal of good quality is found at places all along the western face of the mountain, and on Alpine Mountain and Pilot Mountain. A dozen or more banks have been opened. I visited one place on the brow of the mountain where the coal showed a seam four feet thick. It is an excellent block coal but inaccessible. Putnam, Overton and Fentress all have a supply of coal that will last many centuries, even with means of transportation and a vigorous development. At present there is very little demand for it except by blacksmiths, who usually dig and haul their own supplies.

Lead.—Several beautiful specimens of galena were presented to me which were said to have been found in Clay county, and to have been taken from a large deposit. How much there may be I cannot say, as the person who professed to know where it is refused to impart any information as to the locality of the deposit.

Whetstone Grit.—A fine supply of this is found on Alpine Mountain. It has been worked to some extent and the whetstones are highly esteemed. Coarse and fine grit both occur.

Mineral Waters.—Seven miles southwest of Livingston is a sulphur spring of rare merit. It is said to be of nearly the same quality and produce the same effects as the Red Boiling spring of Macon county.

Chalybeate springs are quite abundant. The most noted spring is on Alpine Mountain, five hundred feet above the valley. This place was once improved and was a pleasant summer resort.

Water Powers—We have given a description of Waterloo Falls on pages 30 and 31, as illustrative of the general character of the streams of this region. There are at least thirty streams in the counties particularly described in this report which supply good water privileges. The best are Spring creek, Roaring river, West Fork, East Fork,

6

Eagle creek (with very valuable falls), Mill creek, Flat creek, Blackburn's Fork and its tributary, Bowerman's branch, Flynn's creek, Calf-killer and its various tributaries, and many others. The whole region has an interpenetrating network of streams capable of propelling water wheels enough to do all the manufacturing of the South. [See White and Warren counties in Appendix.]

Timber and the Lumbering Trade.—No portion of the State has a greater amount of valuable timber, and from it by far the largest supplies are drawn for the Nashville market. It takes about four days to carry a raft from Celina to Nashville, a distance of about 220 miles by water. Raftsmen charge from fifteen to twenty cents per hundred for rafting down after the raft is made up. Usually, however, the head raftsman is paid from $20 to $25, payable when the raft is landed at Nashville. Some six or eight other hands are employed, to whom $10 each are paid. Rafts contain from 200 to 300 logs. Coal boats from Pulaski county, Kentucky, are carried down on the same terms.

The rafts which come out of Obey's river are small, consisting of only thirty or forty logs. At Celina, at the mouth of Obey's river, six or eight of these small rafts are united and a crew obtained for floating to Nashville. The price of trees vary from 50 cents to $1.50 across the stump, the price depending upon the distance from the river and the kind of trees sold. Walnut timber brings about double these prices. Timber trees within three or four miles of the river are getting scarce. The woods have been picked over until really first-class logs are hard to get within that distance, and those that are left are in inaccessible places.

The price of walnut logs ready for rafting, on East Fork of Obey's river, is 90 cents per log, twelve feet long and from twenty up to thirty-six inches in diameter. Extra large logs are worth $1. At Nashville these logs, up to thirty inches in diameter, bring per hundred feet $1.75 ;

from thirty inches up, from $2 to $2.50 per hundred. In buying logs one-third is allowed for squaring and one-fifth for cut of saw. No poplar log under thirty inches in diameter is wanted by the Nashville saw-mills. Over thirty inches $1.25 per hundred is paid, the number of feet being estimated by the rule given above.

In the Nashville market poplar, walnut and ash are in greatest demand, though three-quarters of the lumber used is poplar. The largest supplies are derived from Jackson and Clay counties, but a great deal comes from Fentress and Overton by means of Obey's river. From the region of the Caney Fork a large quantity is rafted down. Between 13,000,000 and 15,000,000 feet are annually floated down the Cumberland. The walnut is mostly shipped in logs to Chicago, Cincinnati, Boston, Philadelphia, and some to Baltimore. The poplar is mainly manufactured in Nashville and shipped to the surrounding country. It is a conceded fact that there is no building timber superior to the Tennessee poplar. It is light, strong, durable and easy to work. A roof made of drawn yellow poplar shingles will last for thirty years.

The price paid for poplar logs is from $5.00 to $10.00 per M. Walnut, as has been stated, brings from $17.00, to $25 per M., Scribner's measure. Squared and delivered on the railroad it brings from $35 to $40 per M, depending upon the quality. About 1,250,000 feet per annum will cover the amount brought to the Nashville market. Usually Tennessee walnut is inferior to the Northern grown walnut; the grain is coarser, and there is wanting, for the most part, the beautiful curls and shadings that make the Northern grown walnut so valuable for veneering purposes.

At the saw mills in Overton and Clay Counties poplar and oak lumber can be bought for $10 to $12 per M. Split staves (barrels) can be had at $12 per M; sawed $4 per M.

Pine lumber is scarce. The pine trees are only found on

the mountain side, generally in places difficult to reach with a wagon.

Labor is very cheap. Good hands for lumbering may be had for for $8 to $10 per month, with board. Without board $12 to $15. Carpenters are worth $2 per day and board. Board can be had at $1.50 to $2 per week in the country; in towns $3. From the oil wells on Spring Creek to Butler's Landing, or Gainsboro, hauling is done at the rate of 35 to 40 cents per hundred.

I cannot close this part of this report better than by inserting the following letter from Hon. J. D. Goodpasture, of Overton county, whose loyal devotion to his county, and close habits of observation, make his suggestions of peculiar value. There is no portion of the State where immigrants can find cheaper or better homes, among a people, too, noted for their hospitality, their integrity, and for their high principles of public honor and duty.

J. B. Killebrew, Commissioner.

OVERTON COUNTY prior to the formation of Clay county, had a voting population of about 2,300, and was at least in point of soil and other natural advantages, an average county of the State. By the formation of Clay, about one fourth of her territory, and one third of her wealth was taken of. She is a stock raising county, and must continue to be so as long as she is cut off from all means of transportation, no other branches of industry will pay. In 1860 she sent South ten thousand head of fat hogs, in 1861 it is said the number was still larger.

She is well adapted to the growth of grasses, and for that reason the raising of horses, mules, and cattle and sheep pays well, but heretofore she has realized more from the hog than other stock. The industrious farmer as a general thing is now doing well, but we labor under many disadvantages. In the first place this was a border county during the war, and for several years after the war, society was very much disorganized. In the next place we are without railroads. These causes induced a large number of our most intelligent and enterprising citizens to leave the county, and those who remained were disheartened, and seemed to have lost to a great extent their energy.

Another great drawback on our farmers: They own too much land and too little money. The consequence is they have been unable to keep up and improve their farms We have had but little immigration to this county. Our people have been unable to sell their lands. Unfortunately those who do not own land are too poor to buy, and a very large proportion of our population are in that fix. What a country needs most is for those who cultivate the soil to own it. No country can prosper where a very large proportion of its population are renters.

In this county as a general thing, the land owners have but little money. They cannot vest much money in hired labor, and for that reason they only hire during the crop season and pay part money, and the balance in provisions. The farmer generally boards his laborers, and pays them about ten dollars per month for their work. By the time the crop is raised, the hired man finds himself in debt.

From the time the crop is laid by till the next spring the laborer is out of employment, usually going from place to place working a day here, and another there for a little meat or corn, and getting as much on credit as he can, which he seldom pays, whilst his family are doing nothing at home for the reason they have nothing to do, and living on the most scanty allowance. I would here state however, that that I believe our people would all work the year round if they could get employment, and get pay for their work.

There is another class of our people quite numerous, owning no land but relying upon renting every year. They generally raise nothing but corn and oats, sometimes a little wheat, paying the landlord one third part of wheat raised. They usually begin their crop about the middle of April, and work till it is laid by, say about first of July, the balance of the year they do nothing or but little. The renting system is the worst of all, both for landlord and tenant. Usually the landlord furnishes the renter with a house to live in and a patch of the best land on the place for garden and truck patches free of charge. The renter as a general thing has an old poor horse or mare frequently blind, (they have a friendship for blind horses,) and bull tongue plow with which to raise his crop. With this horse and plow he scratches the ground and plants his crop. If it is hillside land the soil generally washes away in a year or two. A crop for one man at gathering time is usually about one hundred and fifty bushels of corn (the stock either eat up the oats or they are spoiled by the rust so there are none to harvest) so the landlord gets fifty bushels of corn, generally of a sorry quality, and the renter one hundred. At gathering time the renter generally owes fifty bushels of his corn, for corn he borrowed to raise the crop. Now in order to do any good there must be a change.

What we most need is capital, and more enterprising farmers.

Our farms must be divided, and more of our citizens, must own land, and instead of working three months in the year, they must be employed the year round. We must raise tobacco or something else that will keep our people employed, and that employment must be profitable or they will not work. A renter will not improve another man's land, unless he is paid for it. If he works he must be paid.

If we had railroads and a market for our potatoes, onions and many other things, hundreds would work and raise these things, for market that now do nothing, but we have no railroad, so we must do something else.

Here I would say I believe this is the best undeveloped country in the United States. I believe that the same quality of land, can be bought here for less money than it can be any where else in the Union.

In this county there is fully two thirds of the land that has never been cleared, still in the woods, and of this full three-fourths is excellent farming lands.

There are large quantities of good farming land in the woods, not a stick amiss that can be bought at from $2.50 to $5.00 per acre, and on almost every tract never failing springs.

Respectfully yours,

J. D. GOODPASTURE.

APPENDIX.

APPENDIX.

In my investigations of the oil resources of the State, I visited several counties where no valuable indications of oil presented themselves. A good many objects of interest, however, were brought to my attention, a knowledge of which may prove of value to those desiring to make their home in the State. Among other counties visited were White and Warren, two counties which present a great variety of attractions.

WHITE COUNTY.

This county lies immediately south of Putnam, and contains about 350 square miles. Of this area about one-fourth is mountain land, one-fourth valley, and the remainder bold hills, elevated plateaus, and swelling knolls. The surface varies from high mountains on the east to gently rolling hills on the west, with isolated points that rise up about half way the height of the mountain.

There are three principal topographical features, viz: The Table-land or Cumberland mountains on the east, the Valley lands and the Barrens. The Table-land has an elevation of 2000 feet above the sea. It has a level or gently rolling surface, cut in places by deep gulfs or gorges; pure mountain air, delicious water, and beautiful and sublime scenery. The description of its topographical features, as given in the Resources of Tennessee, are so accurate, that I can do no better than to adopt it.

The mountain slopes on the face of the Table-land and its spurs and outlying ridges occupy a considerable part of the area of the county. The surface on these slopes is for the most part broken and rugged, with many bold cliffs and deep ravines. The escarpment of the Table-land is marked by a line of hard sandstone and conglomerate cliffs, in many places towering high above the tall trees on the slopes below. From the salient angles of these cliffs may be seen extensive and beautiful views of the lower outlying ridges with their intervening valleys and the broad flat and wooded country beyond, extending as far as the eye can reach. At about half the height of the Table-land is the terrace or "bench." This terrace has the same elevation as the tables or tops of most of the little

mountains or outliers. It affords sites for some beautiful farms and orchards, where all varieties of fruit common to the country are produced.
The valley of Lost creek, cut off and completely encompassed by Pine
Mountain, an arm of the Cumberland, is on a level with the terrace.
This terrace was doubtless originally much more extensive than at present, and there are evidences that it covered more than half the area of the
county, including the whole valley of the Calf Killer river and all the
smaller valleys and coves in the county, and also the range of smaller
mountains to the west. By far the greater part has been removed by the
agency of water, but the spurs and outliers are left to tell the tale of its
former extent. The escarpment of the terrace, as it now is, is very much
scalloped by coves and protuberances of large size extend outward, forming
spurs, some of which spread out into plateaus, separated by coves and
valleys. Some of these spurs are cut off by gaps, forming separate mountains; but all, with two exceptions, have a common elevation. The two
exceptions are Pine mountain, between Lost creek and Hickory valley,
and Milksick mountain, west of Hickory valley, both of which are equal
in height to the Cumberland Table-land. A belt of these little mountains,
averaging three miles wide, extends all along the western base of the
Table-land. Interspersed among them are many coves and small valleys.
Separated from these by the broad valley of the Calf Killer, there is a distinct range broken into three parts by large gaps. This range begins with
a spur of Cumberland mountain in Putnam county, which extends first
westward and then southwest around the head of Calf Killer river. The
extremity of this spur is in White county. In a line with it the range of
small mountains extends southwest entirely across White county, termi-
nating near Rock Island in the Caney Fork. This range is cut off from
the spur by the valley of Cherry creek. It is divided by three gaps into
four separate mountains, each of which has a distinctive name. These
gaps are on a level with the valleys, and all of them are large enough to
contain farms. They give easy means of outlet to the open country north
and west. The valley of the Calf Killer lies between the belt of little
mountains along the base of the Cumberland and the range last described.
Its head is in the southeast corner of Putnam county. Narrow at first, it
grows wider as it extends toward the southwest, occupying a belt across
the center of the county, and reaching from one extremity to the opposite.
It is twenty-five miles long, and has an average breadth of about four
miles. The surface is generally rolling, and there are no bottoms along
the river. An interesting topographical feature is presented by the sink-
holes, which are very numerous in this valley. These hopper-shaped
cavities vary in size from ten to one hundred yards in diameter. Their
presence indicates the existence of underground caverns, through many of
which flow subterranean streams. In all this region there is no running
water on the surface, except the rivers and large creeks; all of the springs
being swallowed up by the caves. In many of the sink-holes the opening

at the bottom has become closed by stiff clay or some other obstruction, and in such cases a little lake or pond is formed. There are many of these in all parts of this valley, and they are a convenience to the farmers; enabling them with ease to have water in every pasture. Hickory Valley lies between Pine and Milksick mountains in the southern part of the county. It is five miles long with an average breadth of one mile. Its characteristics are similar to those of the Calf Killer Valley, with which it is connected by two gaps at the upper or northern end. Cherry Creek Valley opens into that of Calf Killer above Yankeetown. It is seven miles long and three quarters to one mile wide. The elevated valley of Lost creek has already been mentioned. In it are a number of beautiful farms, where the people dwell retired and caring little for the changes that agitate the world abroad. The waters of the creek linger lovingly in this Arcadian retreat, protracting their stay by many graceful meanders, and then steal away through an underground channel beneath the mountain into the Caney Fork. There are many beautiful little coves snugly ensconsed among the smaller mountains, generally having one or more outlets into the valleys. Beyond the range of mountains which bounds the Calf Killer valley on the west, are the barrens. Most of the surface is level or gently undulating, and all the streams of water are here on the surface.

Rocks, Soils and Timber.—The rocks on the Table-land are sandstone, and consequently this division has a light, sandy soil, well adapted to the production of wild grasses, fruits and garden vegetable, but too thin for grain; tracts of boggy land along the streams, which, when drained, make beautiful meadows; small trees of the hardier kinds, of which post-oak and black-jack are most abundant. This part of the county is sparsely populated, and is now regarded as of little value except as a summer range for cattle. Most of the farmers in the valleys own tracts of the mountain lands, in some cases amounting to thousands of acres, where they have ranches or "cow-pens." The woods are burnt off in February or March, leaving the surface smooth and clean for the growth of the grass, which then springs up beautifully, and after a few warm days, the whole mountain presents the appearance of an unbounded meadow. Wild flowers grow in great profusion and bedeck with gay colors the emerald sea. Thither the cattle are driven from the farms in the valleys, and attended by herdsmen, who allow them to range at will and graze on the rich herbage during the day, but pen them at night. The Mountain Limestone crops out on the slopes above the terrace, and yields, by disintegration, the elements of fertility to the soils in its vicinity. These terrace or "bench" lands are especially valuable for fruit farms. Some of the orchards never fail to produce good crops. They are peculiarly exempt from injury by frost. The tables of the outliers have a cap rock of sandstone, and a soil in a respects similar to that of the Cumberland table-land. Limestone appears again on the lower slopes, and prevails to the base of the mountain. Too rugged for cultivation, these slopes are nevertheless valuable for the great forests of timber they bear. Sugar maple, beech, ash, walnut, buckeye, linden, wild-cherry, and immense yellow poplars are abundant in the forests. In the valleys the soil is generally good, being a dark brown loam, on a subsoil of strong clay, which lies on a bed of Lithostrotion limestone. The subsoil is of a peculiar red color, made so by oxide of iron liberated in the decomposition of masses of ferruginous chert. In some places these cherty masses are scattered loosely over the surface, in nodules or irregular concretions from the size of a pebble to several hundred pounds in weight. These rocks are troublesome in tillage and wearing on implements, but by gradual disintegration they continually add fertilizing elements to the soil. Most of them are highly fossiliferous, and among them it is common to meet with a large coral of a prismoidal form, known to geologists as the *Lithostrotion Canadense.* The richest lands in the country are in the smaller valleys or coves, some of which appear to have been, at a remote period, the beds of small lakes, from which the water has escaped, leaving a deep, rich alluvium, well mixed with sand from the surrounding heights. With good tillage the soil is inexhaustible, and it is very easy of cultivation. When the country was settled, the coves were covered with a very heavy growth of beech, sugar-maple, buckeye and yellow poplar, while an undergrowth of cane-brakes rendered the surveying of the lands a work of great difficulty. In the barrens much of the soil is thin and deficient in lime. Sandstone prevails, valuable for building, but imparting no fertilizing quality to the soil. Much of the surface is level or gently undulating, and thinly wooded. Post-oak, suitable for cross-ties, is abundant. At several places, however, red clay and limestone prevail, and furnish sites for a number of good grain and fruit farms, and the less fertile portions furnish a fine range for sheep and cattle.

MINERALS.

Coal.—The coves which indent the mountain side disclose and render accessible several valuable seams of coal.

Railroads may be constructed with an easy grade, and sufficiently below the coal to permit the construction of chutes for the loading of cars. Several mines have been opened east of Sparta, which supply a most excellent grate coal that burns with the brightness of a pine-wood knot. Four miles east of Sparta I examined and measured a stratum of coal four feet in thickness. This stratum is parted in the center by a seam of cannel coal, which at the outcrop was only about one inch thick, but gradually increased, so that at the distance of twenty feet it measured three and one-half and four inches. This is called the Fisk Bank, and promises well, the slope of the mountain being very gradual, and the coal a hard block, with a semi-lustrous appearance, and will bear transportation well. One hundred yards south of this place the same stratum has been worked and mainly used for blacksmith purposes. Two hundred yards further the same stratum is now worked to a limited extent, the quantity taken from it being only 125 bushels daily. The coal at this place shows an eastward dip, and if worked from the present opening much difficulty will be met with in keeping the mine drained.

The underclay at this place is four or more feet in thickness, while a bluish shale overlies the coal, which is but the underclay to a stratum above. The bed of clay underlying the stratum rests upon sandstone, which is here about 150 feet in thickness, itself resting upon the mountain limestone. At this place three distinct openings have been made all of which display good workable coal.

Little's Bank lies a quarter of a mile further north, which is more extensively worked than any other in this county—nearly all the grates and blacksmith shops in Sparta being supplied from this place. The entry has been driven in about seventy feet with no cross entries. The thickness of the seam at the outcrop is 44 inches, which at the distance named above has increased to four and a half feet. Three feet below this seam is another three and a half feet thick, which was worked many years ago and furnished coal equal in quality to the one now worked. Six feet above the first mentioned, another seam appears two feet thick, furnishing coal of a similar character. The three seams at this point will aggregate a thickness of ten feet, and it is quite probable that they all run into each other.

Various other openings have been made, enough to demonstrate the fact that the whole western escarpment of the Cumberland Table-land in this county has outcrops of coal which, in the aggregate will measure from eight to ten feet in thickness. The distance from the extreme southern opening to the one lying further north, is twenty-two miles, while east of Sparta sixteen miles, and twelve miles from Settle's Bank, at Scarborough's Mill, coal appears in a seam ten feet thick. This coal is found in a valley-like depression in the mountain, and is covered by only about a foot of sandy soil. In general terms we may, in confidence, affirm that the whole eastern part of White county is underlaid with coal of great purity and excellence, easily worked, and covering, in the aggregate, quite one hundred and twenty-five square miles. Nowhere in the State, are the seams more trustworthy or the coal of better quality.

It may be well, in this connection, to state that the same outcrop continues southward in Van Buren county. On Big Hill branch, a wet-weather stream and tributary to Cane creek, six miles east of Spencer, good lump coal occurs in a seam six feet thick. It has been used for twenty-five years and is highly esteemed for its welding properties.

Another presentation of coal occurs in the bed of the same branch, two hundred feet below the top of the mountain. Northwest of this, two miles

on the east side of Cane creek, is Mooneyham's bank, where the seam is
nearly one hundred feet above that last described. On both sides of Cane
creek, for a distance of twelve miles, coal occurs in workable quantities as
also in the cross ravines. The portion of our coal field in White and Van
Buren counties has been but little noticed, from the fact that no means at
present exist for transporting the coal to market.

I am clearly of opinion, after having visited nearly all the mines in
the State at present worked, that, in quantity and quality, no coal in the
State is superior to that found in the two counties under consideration.
It belongs to the lower coal measures, all the seams being below the conglomerate except those on Clifty creek and at Scarborough's mill. The
seams are trustworthy, the coal is of great purity with but a small percent-
age of sulphur and free from slate. It burns freely, leaving as a residuum
a white ash. It is much harder than the Sewanee or Rockwood coals, and
unlike them, has not a shelly or crushed appearance, and will bear transportation as well as the Battle creek or Upper Cumberland coals. In no
place is such a body of coal found in the lower coal measures.

The following is a section taken at Little's Bank, which will no doubt
apply to all the western margin for several miles :
Beginning at the top and descending we have :

1. Conglomerate..40 to 60 feet.
2. Shale..75 feet.
3. Sandstone...12 feet.
4. Coal... 2 feet.
5. Shale and Fire Clay.. 6 feet.
6. Coal, main entry 44 inches at out-crop, cubical............. 4½ feet.
7. Fire Clay and Shale.. 3 feet.
8. Coal worked for many years... 3½ feet.
9. Under-clay.. 2 feet.
10. Space down to Limestone, Shale and Sandston·............50 feet.
11. Mountain Limestone.

A peculiarity in the lower coal measures in this region is noticeable
In Pennsylvania these are called the "Barren Coal Measures," because the
coal lies in pockets and is not reliable. In the counties of Marion and
Hamilton, though the seams below the conglomerate, never run out, yet,
they are liable to continual variations in thickness. Sometimes the
coal lies in great masses eight or ten feet thick, and then thins out to a few
inches.

At the Ætna mines, in Marion county, a seam was opened which at the
outcrop was six feet thick, increased to nine, and then fell off to three
thus showing the lenticular character of the strata of coal in the lower
measures in that county. In White, on the contrary, the seams so far

have varied very little, though increasing in thickness gradually with the length of the adit.

Iron Ore.—While the coal supply of White is in excess of any probable demand for a century to come, the iron ore, though good, is more limited in extent. A little west of north from Sparta, and at a distance of three and a half miles, is a series of low hills, in which pot ore (limonite) abounds. The surface of these hills is covered with a water-worn gravel of a dingy yellow color, intermingled with masses of chert and yellow clay. The ore occurs in masses of all sizes, up to blocks which would weigh three hundred pounds. One lump which was dug out for me weighs over two hundred pounds. The hills in which the ore occurs are conical in shape and about thirty feet above the general surface of the Calf Killer valley. They cover from one to five acres, and are separated from each other by gentle depressions. The characteristic growth is scrubby black jacks, post-oaks and hickory. The iron deposits extend for the distance of four or five miles in a northeasterly direction.

The Board Valley Mountain, an outlier of the Cumberland table-land, lies ten miles north of Sparta. It is an elevated ridge about 500 feet high, eight miles long, one mile wide, and bears northeast and southwest. The southeastern side of this mountain is sandstone, while limestone is the prevailing rock of the northwestern side. On the top and southeastern slope brown hematite occurs in considerable abundance, enough from the external indications to supply a twelve-ton furnace for many years. Workable ore is found throughout its entire length. The mountain is heavily timbered on both sides with poplar, oak, walnut, chestnut and other varieties.

As early as 1825 a forge was in operation near Sparta, which made nearly all the bar iron consumed in the county. The ore used in this furnace was taken from the banks $3\frac{1}{2}$ miles north of Sparta. The iron was highly prized for its toughness, strength and hardness, and was largely used in the manufacture of plows and horse shoes. For the latter purpose, owing to its great hardness, it was considered especially valuable.

WATER POWER.

Not the least valuable of the undeveloped resources of the county is the large number of streams that supply water-power. The rapidity with which these streams descend from their elevated sources gives them a wonderful momentum. The estimated power of these streams are sufficient to work up all the cotton grown in the United States.

How Water-Powers are Estimated.—It may not be amiss to observe that the dynamical force of a stream is found by the formula $58.23\,s\,v^3$, in which s is the cross section of the fluid current in square feet, and v the velocity in feet per second—so that if the velocity of the current be 10 feet per second, the mean depth 2 feet, and the mean width 15 feet, we shall have

$$s = 2 \times 15 = 30$$

$$v^3 = 10 \times 10 \times 10 = 1000$$

Now substitute in the formula

$$58.23 \, s \, v^3$$

and we have

$$58.23 \times 30 \times 1000 = 1746900.00.$$

Leaving off the two decimals the remaining figures represent the dynamic force of a current. Now, to get the horse-power, divide the dynamical force by 33000, which is the unit of measure, and we shall have for the supposed stream 52.94, or nearly 53-horse power.

Reduced to rule, we may say, to obtain the number of horse-powers in a stream—

Measure the velocity of the stream in feet for one second; measure also in feet the mean width and the mean depth of the stream, then multiply the width by the depth and this by the cube of the velocity and the result by 58.23. Cut off the two right hand figures and divide by 33,000, and the quotient will be the dynamical force of the stream in horse power.

Of all the water powers in the county, that on Caney Fork at the falls, if not the best, at least is the most powerful. This stream is one of the largest tributaries of the Upper Cumberland. Taking its rise on the Cumberland Mountain about eighteen miles east of Sparta it descends through a deep, dark, narrow gorge, hemmed in by frowning cliffs for twelve or fifteen miles, when it debouches into an undulatory valley plain. Passing westward through this valley by many winding ways, it plunges, at Rock Island, over a siliceous limestone, by a succession of falls and rapids, for two and a half miles.

At Rock Island, where the piers for the railroad bridge have been partly built, there is a fall of five feet, which might be increased by a dam of any required heighth. From the island to the principal falls, a distance of about two miles, there is a fall of five feet. At this point the water descends perpendicularly twenty-five feet. Below the main fall, for 250 yards, there are rapids with a fall of six feet, when there occur three successive falls within one hundred yards, each of about twenty feet. Then succeed rapids for thirty yards, with a fall of six feet. Below the rapids the water is eddy for 150 yards. Below the eddy water there are rapids for a hundred yards or more, with a fall of six feet. From this point to the principal falls, a distance of a quarter of a mile, the aggregate descent, as measured by Major Falconett, civil engineer, is 96 feet. Below the rapids last mentioned there is a succession of shoals, until at the distance of three miles the Horse Shoe falls occur, where there is a perpendicular descent of six feet. The current, from this place to Bailiff's mills, a distance of two miles, is rapid.

Four miles below, at Frank's ferry, is the head of steamboat navigation. At many places in the river the channel is compressed within a space of twenty yards, while at others it widens to one hundred yards or more. The average breadth of the stream is about seventy-five yards.

The banks and bottom of this stream from Rock island to the foot of the rapids are composed, as before remarked, of ledges of hard siliceous rock which have withstood the erosion, while the softer rocks below, mainly the Trenton limestone and shale, have not been able to resist the continued corroding action of the water.

Dams, by reason of such banks, can be made durable without any danger from undermining or a diversion of the stream by pressure around the ends. Material for their construction is cheap and convenient. The lay of the land for the erection of buildings is not good; high, overhanging bluffs or steep banks extending nearly all the way. Nevertheless, a few good sites are found under the bluffs, and the water could be conveyed through artificial flumes below the falls two or three hundred yards, where the bluff gradually subsides into a comparative level surface.

It would be difficult to estimate the great power which could be developed at this place. It would come within the range of possibility to say that, throughout the entire distance from the foot of the rapids to Rock island, a distance of two miles and a half, a dam eight feet high could be constructed on an average for every four hundred yards, so as to secure a storage of at least 20,000,000 cubic feet of water. From the island to the last rapids, the velocity of the current would average not less than eight feet per second, while a section of the river in ordinary water below the island and above the principal falls, would give near 900 square feet. Upon this conjecture each dam would represent a 760 gross horse power, or the stream for the distance under consideration would furnish power equal to 9120 horses.

Yet the Caney Fork is by no means the most available water power in the county. Many others have better sites for the building of mills, and which will furnish power enough to drive any ordinary mill or factory. One of these is Falling Water, a tributary of Caney Fork. The stream is not one-eighth as large as Caney Fork, but it is very valuable. At Williams' mill, twelve mile from its mouth, there is succession of rapids where the descent is 200 feet in 700 yards. The volume of water is sufficiently large for any manufacturing purposes. At this mill there is a present power of about 80 horse, and this could be multiplied by the construction of dams several times within half a mile. The mill at this point has no other dam than a log pinned down in the bed of the stream, which diverts the current into the mill race.

Below the mill, at the site of a former mill, there is a natural rock flume, into which all the water at ordinary stages is gathered, and was made to run the mill without any dam whatever. The supply of water of this stream is constant. Owing to the rapid fall the water never gets too high, and an experience of fifty years shows the remarkable fact that no mill has ever been stopped by excessive freshets.

The cascade falls, half a mile below this, is one of the most picturesque in the State. The stream at this point has cut down through the silicious limestone seventy feet, and through a bed of black shale thirty feet in thickness, and carved out a deep channel in the Nashville limestone. The bluffs on each side rise up quite 250 feet, and the water plunges over two faces of an angular rock in a perpendicular fall of 120 feet. Just before reaching the bottom it strikes against a shelving mass of rocks, which lashes the water into great horizontal cylinders of spray. A beautiful rainbow in the evening, when the sun is shining, rests upon the surging mass of waters. During the winter months the spray congeals upon the tops of trees in the gulf below, and accumulates in such masses as to break off the tops.

Taylor's creek, a tributary of Falling Water, passes through the north western part of the county, and though the volume of water is small, it supplies excellent power easily and cheaply utilized. At Fancher's mill, two dams are erected within a distance of two hundred yards. The fall of the stream in a distance of 300 yards is 69 feet.

Below the lower dam there are rapids for 200 yards, then a succession of falls ten, twenty and one hundred and sixteen feet. The stream has worn down a channel in the solid rock fifty feet or more, before reaching the main fall, so that the falls are not so high as the bluff by that distance. At the falls the bluffs widen out so as to make a semi-elliptical grotto or cul-de-sac. This deep grotto or chasm extends down to Fallingwater, five or six miles distant, and the scenery is inexpressibly wild and picturesque.

Town creek, a tributary of the Calf Killer, furnishes admirable water-power. Nearly the entire volume of water is furnished by a spring two miles west of Sparta. The stream flows in a southeasterly direction, and the entire length is only one mile and three-quarters. Yet there are several mills upon it, and the supply of water varies very little in summer or winter.

The falls of the Calf Killer river supplies a very large amount of water-power within one mile of Sparta. Upon these falls the Sparta factory was erected many years since, and up to the breaking out of the war cotton and woolen goods were manufactured in large quantities, and the factory supplied remunerative employment to a great many women and children. The machinery and looms were shipped south during the war, and the building has not been re-stocked since. These buildings consist of several tenement houses and a large brick, sixty by one hundred feet, four stories high. The walls, roof and floors of the factory building are in a good state of preservation. The dam has been destroyed, and trees are growing up in the race. The river at this point has a very rapid descent, a succession of falls occurring for nearly a mile, over an indurated sandstone. At the factory there is a fall of fifteen feet within 300 yards. When the factory was in operation there was a seven-foot dam above the falls, which gave a head of 22 feet at the factory. The height of this dam might be increased to ten feet.

The hard sandstone [which forms the bed and banks of the stream at this point resists the erosion of the water, and is very favorable for the construction of water-tight dams. There are many other points on the Calf Killer where excellent mill and factory sites can be procured. Several good mills are already in successful operation, and it is to be regretted that more of its excellent and available force is not utilized in spining and weaving our cotton and wool into textile fabrics.

ROCKS OF COMMERCIAL VALUE.

The building stones constitute one of the important resources of the county. The limestones, of which there is a great variety, are hard, compact and durable. Hydraulic rock occurs in three varieties, blue, white and grey, and is found in abundance, cropping out all along the Calf Killer and upon the sides of the mountains. A limestone of a clouded white appearance occurs in great quantities, and has been quarried and wrought into tomb-stones. Whetstone quarries have been opened on the western slope of the Table-land. The most noted quarries are on the right and left of the road leading from Sparta to Kingston, and another on Lost creek. The whetstones obtained from these points are equal to the best in market. The sandstone heretofore spoken of as occurring near the Sparta factory, has a sharp grit, but is not equal to that obtained from the quarries mentioned.

Flag-stones of an excellent variety are found in the low spurs and foot hills of the Cumberland mountains. They vary in thickness from half an inch to ten inches; many of them have a surface as smooth as if they had been dressed, while others are ripple-marked. The stones form the cap rock of the hills, and lie above the limestone. A quarry resembles a pile of planks of varying thickness. Each stratum maintains the same thickness throughout. These stones are easily quarried, and they could be got out with mechanical power sufficient, large enough to cover an acre in extent. They are used for pavements, and some of them have even been used for hall floors. If river or rail transportation was afforded, they would bear transportation a great distance, as the cost of quarrying them will not exceed one-fifth the cost of sawing and dressing limestone. Potter's clay of a good quality is found in the northwestern part of the county. It is extensively worked.

Salt Wells.—As early as 1818, wells were sunk on the Calf Killer river three and a half miles northeast of Sparta. They were re-opened during the recent civil war, and supplied the surrounding country with salt. Carburetted hydrogen gas issues from the wells, which burns with a brilliant white light. The supply of this gas is continuous, and it has been known to burn without intermission for a period of six months, making a light of such refulgence that persons at the distance of several miles could see to read large print distinctly. The water which issues from these wells is of a deep blue color, and is brought up from a depth of three and four hundred feet. It has a sulphurous and brackish taste. Fifty bushels of salt per day have been made from these wells by the evaporation of the water in shallow kettles. The timber for miles around has been cut for fuel to be used in evaporating the water.

East of Sparta, and on the road leading from Sparta to Bon Air Springs, a stratum of earth appears which resembles a mass of disintegrated shale. This lies above the whetstone grit, and between it and magnesian limestone. There are three varieties of it, differing mainly in coloring matter. The lowest bed is blue, not unlike the underclay of the coal measures; the second has a purple color, and the uppermost stratum a dark reddish color. This last lies immediately under the limestone, and is said to have been used with good result as a fertilizer. The bed is five feet thick, and could be used to great advantage on the sandstone soils of the mountain.

Mineral Springs.—There are several mineral springs in this county some of which are noted for their health-giving properties. The most noted of these is Bon Air, which is 1827 feet above the sea level. The water is chalybeate, and many years ago it was a place of summer resort. But the buildings have all gone to decay. The scenery from this point westward is varied, extensive and grand. Rounded peaks, long ranges of swelling heights, deep gulfs of verdure along which the Calf Killer and its tributaries flow; cultivated fields, sometimes running high up the slopes of the mountains, scattered farm-houses—these form the principal features of the landscape. The high elevation of this place rising as it does far above the malarious atmosphere of the bottoms, assures vigorous health to the sojourner.

Thus it will be seen that White county has immense capabilities, and the railroad which was projected to run to Sparta from McMinnville, and which has been graded, would offer a wide field for the industrious. Cheap water power and rocks of great commercial value abound; also soils of more than average fertility, well adapted to the growth of tobacco, corn, wheat, oats, hay, potatoes, and upon which are grown excellent cotton; these will insure a brilliant future for the county. Good roads are wanting. The schools are improving, and the public mind is turned to the advantages which would result from a utilization of the great natural forces and agencies which abound in the county.

FARMING INTERESTS.

With a good soil, and in a region where droughths seldom occur, on account of the proximity of the mountains, the farmers of White county ought to be prosperous. The negroes, to a large extent, have left the county, and the supply of labor is scarce. The wages of farm hands vary from $8 to $12 per month with board. The principal crops, named in order of their importance, are corn, wheat cotton, oats, sweet potatoes, Irish potatoes, tobacco, rye and turnips. Dried fruit is an important article of trade, as also eggs and chickens. At one country store I found the following shipments of country produce:

Dried apples..30,000 pounds.
Eggs..100 dozen per week.
Feathers..1,200 pounds per year.
Dried blackberries...5,391 pounds.
Ginseng..300 pounds.
Chickens..1,500 per year
Cotton in seed.. 15,000 to 30,000 pounds per year.

In addition to these, the staple products are bacon, wheat, peas, beans, peaches and a small quantity of tobacco. The cultivation of this last crop is yearly increasing. The farmers, as a class, are very independent, and make upon their farms nearly everything they live upon, saving always a surplus sufficient to buy sugar, coffee and salt. Home manufacture is carried on extensively including jeans, linsey blankets, carpets, matting, cotton and woolen socks, cotton cloth, flax, linen, baskets, shuck collars and ropes.

Farms are usually small and well cultivated. There is a great diversity of crops, and in no portion of the State does industry on the farm pay better. Upon the best lands it is not uncommon to gather 50 to 75 bushels of corn to the acre. Good farms are worth from $15 to $40 per acre. Mountain lands from 50 cents to $1.00. Barren lands from $1 to $3 per acre. The best farming lands are in Calf Killer valley.

As a fruit region portions of White county are unsurpassed. Mr. D. S. England, living five miles from Sparta, has over 40 acres in orchard. Four acres are devoted to pears, of which he has a great variety. He cultivates 40 varieties of peaches and 150 of apples. Rarely does he fail to have a good crop. His farm is the only place where I have seen the Scuppernong grape do well upon a clayey soil. The vines upon his place bear full every year. He has other varieties, and manufactures considerable quantities of wine every year.

Farm Stock.—There is a general disposition to improve stock of all kinds. Numerous importations of fine hogs and bulls have been made within the past few years, and the impression is becoming general that poor stock does not pay.

Evidences are not wanting to show that this region, in the dim and shadowy past, was occupied by a singular race of human beings. Extensive grave-yards are found scattered over the valley of the Calf Killer near the bases of adjacent mountains. Skeletons are exhumed from 23 to 26 inches in length. They are usually found buried in a sitting posture, from one to three feet beneath the surface. The skeleton is always found in an oblong vault lined with flag-stones. In every grave is found an earthen pot containing beads made of shells or stone, and sometimes of dark earth.

WARREN COUNTY.

Excluding the portion of the Cumberland Table-land, Warren county may be said to be flat highlands, but sufficiently cut by streams, with tolerably deep valleys, to give contrast and variety to the surface. The eastern portion is made rough by the spurs and outliers of the Table-land, and supplies many mountain valleys, coves and often wild, picturesque gorges, precipices and water-falls. The southeastern part of the county lies on the Cumberland plateau, and has the elevation, soil and physical features which pertain to that region. Three-fourths of Warren county rests upon a bed of red clay, which has the capacity of retaining moisture without suffering from a want of drainage. Calcareous and siliceous, they combine the strength of the one with the friableness of the other. The remainder of the land is mountainous, but some of the best lands are found in the coves. These are usually very productive, and yield from thirty to forty bushels of corn per acre, while for fruit they are considered unequalled, especially for the apple.

There are but few counties in the State presenting more attractions than the county of Warren. Its bold topographical features and beautiful scenery, healthy climate and fertile soil, its grand supply of water-power and the value and extent of its mineral wealth, all assure for it a bright career in the future.

Its water-powers and manufacturing industries deserve special mention. The principal stream is Collins river, which rises on the ragged edge of Grundy County, and gathering strength as it descends northward to the undulating plain below, pours its flood into Caney Fork, just above the great falls already mentioned in the description of White county. Its length is twenty-five miles, its average width about 250 feet, and depth two feet. It is a quiet stream, for the first ten or twelve miles after leaving the base of the Cumberland mountains falling about three feet to the mile. Toward its mouth its fall increases to eight or ten feet per mile, and the stream becomes turbulent, rushing between rocky banks, composed of hard siliceous limestone. There is very little bottom land in its lower course, and mill-dams with ten feet fall could be erected every mile or two. The Southwestern railroad from McMinnville to Sparta is graded on a line generally parallel to the river, and within a short distance of it.

Barren Fork, the tributary of Collins river, is fifteen miles long, has a volume of water about two-thirds of that of Collins river, but is much more constant in its supply of water, rising as it does in the "flat woods" of Cannon county. All streams rising in the "flat woods" find a more regular supply of water than those rising on the mountains. This stream is never affected by drought to such a degree as to suspend operations of the mills upon its banks.

Hickory creek, a tributary of the latter, has about one-fourth the capacity of Barren Fork. and is more fluctuating in its supply of water. Nevertheless, it furnishes some very desirable water-power and within the distance of six miles turns the wheels of several excellent mills. This stream drains the most desirable farming district in the county, the soil being a clayey loam, based upon an unctuous ferruginous clay that retains in a remarkable degree all fertilizers put upon it.

Mountain creek and Charley's creek run nearly parallel, and are tributaries of Collins river. Both are excellent streams for milling purposes. Charley's creek falls about 20 feet per mile for the first four miles above its mouth. Near its mouth there are two woolen mills, a flouring mill and a saw-mill, all within the distance of half a mile. Mountain creek has several excellent mills. These streams rise in the "flat lands" of Cannon county, and though having only one-fifth the capacity of Collins river, they never fail to supply a sufficiency of water for the mills on their banks. For every half-mile, for the distance of four miles above the mouth of Charley's creek, a dam giving eight feet fall may be constructed.

Rocky river, in the eastern border of the county, rises among the mountains of Van Buren county, and runs northwest, emptying into Caney Fork at Rock Island near the mouth of Collins river. Being a mountain stream, it is very variable in its supply of water. The great falls of Caney Fork has a fall of ninety-six feet within the distance of one quarter of a mile. This water-power, though immense in its force, is very inaccessible, the stream being hemmed in by precipitous bluffs, rising to the height at the falls of about 100 feet. The power, however, could be transmitted by wire belts to any desirable distance. This is now done on such streams in Scotland with a loss of only ten per cent. of power

MANUFACTORIES.

The manufacturing interests of Warren county are considerable. The Annie cotton factory, owned by Asa Faulkner & Son, is situated upon Barren Fork, within half a mile of McMinnville. The amount of capital invested here is $85,000; number of operators employed, 60; average wages paid each hand per day, 50 cents; amount of cotton consumed in twelve months, 240,000 pounds; number of yards produced of 4—4 sheeting, 710,000; number of looms, 60; number of spindles, 2,016. There is a fall in the stream at the factory which gives a head of thirteen feet within one and a half miles, furnishing a sixty-horse power for use at the factory. The principal market for these goods is Louisville.

The McMinnville woolen mills, owned by Faulkner Brothers & Cantrell, three miles north of McMinnville, employ 39 operators. They are paid the following prices:

20 Weavers, 1½ cents per yard.
4 Spinners, 40 cents per day.
4 Corders, 75 and 40 cents per day.
2 Watchmen, 60 cents each per day.
2 Drummers $1.00 and $1.25 per day.
2 Extra work hands, 40 cents per day.
2 Extra mixing, picking hands, 60 cents per day.
2 Foreman (owners of the mills.)
1 Book-keeper, $1.75 per day.

All are natives and white, and are said to be unsurpassed for efficiency and constancy. Tramp labor has proved a failure.

Capital invested in the mills $30,000.

Amount of wool used in 12 months, 35,000 pounds.

Number of looms 20.

Value of goods manufactured per annum $50,000.

About 10,000 pounds of wool are obtained in the county, remainder mainly from Smith and Wilson. Southdown and Cotswold wools are preferred, but these are more difficult to work than the short staple of the mountain sheep. The jeans, 7, 9, 10 ounces, made at these mills, find a ready sale in Smith, Wilson, DeKalb, Coffee and Warren counties. So great was the demand last winter that the proprietors doubled the capacity of their mills. Everything about these mill shows a growing thrift and prosperity. The owners are the chief operators, and to this fact is due the rapid growth and constant expansion of their business.

The Hub and Spoke factory is located at McMinnville, and owned by T. F. Burrough & Co. The amount of capital invested is $30,000; number of hands employed, 40; rate of wages paid is from 50 cents to $5 per day; value of annual products $60,000. This establishment makes handles, buggies and wagon material, except hubs and bent work; ships to to all parts of the United States, Germany, England, and some pick handles to Australia.

LANDS, SOILS AND CROPS.

The lands for the most part being situated on the Lithostrotion bed, have the characteristic chocolate color, and are naturally very fertile. In some respects these lands are to be preferred to the rich black lands of the Central Basin. They have the capacity of resisting a drought much longer. The best farming lands are on the old Winchester road, on Hickory creek. The soil is a deep red clay, and was originally covered with hickory. It wears well. The farms are generally small.

Wheat.—Boughton wheat and Lancaster red are the most generally grown. Lancaster red is thought to yield the largest quantity per acre.

Wheat is generally sown broad-cast on corn lands, and late. The yield is not more than eight bushels per acre, by this process. It can be made to yield 20 bushels per acre. One gentleman who devotes his time mostly to wheat and clover, raises excellent crops of wheat by rotating with clover and sowing early in October. He thinks he will be able to make 30 bushels per acre.

Wheat fills better on Cardwell Mountain than in the lowland; this mountain is very fertile, especially on its nothern slopes. Commingled with the soil are bones and muscle shells. The wheat is shipped to Augusta, Georgia. The cost of shipping from McMinnville to Augusta, is about fifty cents per bushel.

Corn.—White corn is generally planted. A small quantity of yellow corn is planted for stock. Planted about the middle of April upon a soil scratched by a bull-tongue three inches deep, and planted slovenly, plowed about three times, the yield is from three to eight barrels per acre; no surplus is shipped, that being fed to stock.

Oats.—Rust severely affects the oats, and but few are sown.

Sorghum—is extensively cultivated—sells at 20 cents per gallon; generally 100 gallons are made to the acre.

Irish Potatoes.—This vegetable finds a congenial soil. The favorite varieties are the Northern Russets, Peerless and Early Rose. Very few are shipped; not as many as are imported. Very little attention indeed is paid to any vegetable for shipping.

Tobacco.—An effort is making to raise tobacco. It would add greatly to the income of the farmers of the county. It is thought about 100 hogsheads will be made in the county this year.

Meadows.—Herds grass pays better than any other grass grown, though I saw some timothy. In some places blue-grass grows very well, especially on hill sides where limestone comes out and where walnut trees grow. Wherever walnut trees grow blue-grass flourishes with great luxuriance. August is said to be the best time for sowing herds grass. It is put up in stacks in the field. On the flat lands it is a standard crop. Some hogs and a few mules are carried South. Very few beef cattle are sold. Clover grows admirably, but is not raised to the extent it should be; it is generally cut for hay.

Sheep.—Common scrub sheep are for the most part raised, yielding from 2 to 2½ pounds of wool each. The wool is coarse. There is but little demand for the Merino wool; it is too fine to be used in the factories of the county.

Fertilizers.—No commercial fertilizers are used in the county. Stable manure is generally hauled out and applied to the hills of corn.

FRUITS AND ORCHARDS.

No county in the State has paid so much attention to fruit culture as Warren. It is, *par excellence*, the fruit growing county of the State. In that belt of the county overlooked by the mountains and lying at the base, fruit orchards are almost continuous. It is no uncommon thing to see apple orchards embracing fifty, sixty, one hundred, and even three hundred acres. Almost every variety of fruit has been tested, but the following varieties are thought to be best adapted to the county:

Summer Apples.—Early Harvest, a good shipper; Early June, a good shipper, follows the Early Harvest. Horse apple succeeds next, and is good for eating, drying, cooking and brandy making. It is said to make the finest-flavored brandy. Buncomb is a very soon apple, and is chiefly prized for making brandy, requiring only two bushels for a gallon of brandy. The tree is a long liver, and will flourish upon a poor, thin soil, which makes it very valuable for that portion of the county embraced in the "Barrens."

Fall and Winter Apples.—Red Pearmain is good for all purposes. It is an excellent market apple, bearing shipping well. In color it is red, has a fine flavor, and in shape is oblong resembling the Sheep-nose. Winesap is a great favorite, the tree is hardy, long-lived, and bears well. Hall's Seedling is a favorite winter apple. The following list of apples is recommended by a leading orchardist for the valley lands, allowing five acres for a small orchard:

5 Striped and 5 Red June, 5 Early Joes, 5 May apples, 10 common Horse, 10 Sweet (of different kinds), 10 Queens, 10 Gribble, 10 Poplar-Blacks, 10 Winter Horse, 10 Sheep, 10 Hall's Seedling, 10 Winter Sweet 20 Jennett, 20 Newtown Pippin, 10 Fall Pearmain, 5 Turner Grand, 5 Lady Finger, 10 Sweet Limber Twig (can tell them from the common Limber Twig only by the flavor, they will mellow much sooner, but will not keep so long), 50 common Limber Twig, which will make a small orchard full and comp ete.

Apples for Mountain Top.—For the mountain top the following is a list that succeeds well: Wine Sap, Limber Twig, Spotted Buck—this last variety is a fine keeper, it does not do well in coves—Cagle apple does well on the mountains, also the June Red and Horse apple.

The Limber Twig is most admirably adapted to the coves and mountain tops, while in the valley it is to some extent a failure. On the mountain this apple is juicy, tender and brittle, with a blood-red color. In the valley it rots and specks on the trees, and the fruit is tough and leathery with a green color, totally unlike the apple on the mountain, though of the same variety. This variety bears a very heavy crop every alternate year, the year between the trees only bear about one-third of a crop. The apples are kept by putting them up in the field and covering up with leaves. After the severest winter they will come out perfectly sound. The rains do not hurt them, while freezes seem to be a benefit to them ; they will not bear transportation, they are essentially brandy making, and are said to be much better for making brandy after freezing. They are never good until late spring. About two bushels of these apples are required to make one gallon of brandy.

The Vandevere is a good fall apple; there are trees existing in the county fifty years old. The Wine Sap grows to perfection in the clay lands, and is to the valley what the Limber Twig is to the coves and mountain sides. Hall's Seedling grows better in the dry lands than any of the other varieties. The trees are long-lived and require very little attention. It is an excellent fruit for family use, but is too small for market. The Smoky Twig is a late fall apple, but with careful handling can be kept until spring. It has just enough acid about it to prevent it being called a sweet apple. It also grows well in clayey soils. This apple does well and is not liable to spot. Spotted Buck is a good apple but yields sparsely. The Vanderbilt, a late fall apple, on the whole does well, though cannot be said to do so well as many other varieties; it is a never-failing producer, but is liable to speck.

As a general rule an orchard reaches its maximum production from eight to ten years. With care it will not show any evidences of decline for ten years, after which time it gradually declines for fifteen or twenty, then and is cut down or abandoned.

The yield of a good orchard in its prime, is estimated at an average, in a good fruit year, of ten bushels to the tree, on good soil, and there are from fifty to seventy-five trees to the acre. It is no uncommon thing to see from 20,000 to 30,000 bushels of apples put up on an orchard farm to be distilled, or sold during the winter.

There are often partial failures in fruit, resulting from bad seasons or for the want of proper culture. There is but little complaint from disease, the trees are generally healthy ; orchards of a thousands trees are seen and not one tree will be missing. There are no prevailing diseases. A few die from imperfect grafts and diseased roots, made so by compacting together in a tough subsoil.

Brandy-making.—The making of brandy has always been regarded as very profitable, and in certain portions of the county, it is almost the only article of sale.

A very large proportion of fruit is distilled as the apples are taken from the trees. Very often the apples are ground up and put in open vats or tubs, where most of them will remain for six or eight months without impairing their value for making brandy. At the time of my visit, in July, there were distilleries running on apples ground up the preceding November. It is estimated that it takes one-half the value of the fruit to convert it into brandy. The estimates of the profits of a good orchard are about as follows:

One acre containing from 49 to 64 trees averaging from 5 to 10 bushels each, will yield from 300 to 500 bushels, and it takes 2½ bushels winter apples, 3 bushels of fall, and 4 bushels of summer, as a general thing, to make one gallon of brandy. Five hundred bushels of apples, allowing 3 bushels to one gallon, will make 166 gallons of brandy, which, at $1.60 per gallon, will be worth $265.60. From this deduct 90 cents per gallon for revenue, and it will leave a profit of $116.20 per acre. But of this amount one-half should be deducted for labor and fixtures for converting the apples into brandy, so that a clear profit of $58 is made to the acre. But this profit is only realized on orchards in their prime, and in a good fruit year, and when the whole crop is converted into brandy.

In the case of H. L. W. Hill against Meadows, which was adjudicated in the Chancery Court shortly after the war, the affidavits show that 940 gallons of proof brandy were made from an orchard of five acres.

Mr. Jesse Nunley, one of the most extensive fruit growers in the United States, made 1,950 gallons of brandy from 200 trees, or about four acres of orchard. The orchard was in Nunley's cove, 15 miles south of McMinnville. The soil is dark and loamy, with some admixture of mountain sand. It is very friable and very rich.

There are 77 distilleries in Warren county, and from the 1st of August, 1876, to 1st August, 1877, 40,155 gallons of apple brandy were made, or about 1,000 barrels.

Apple Trade.—A very active trade is springing up in apples, many small farmers buying up the apples and carrying them on wagons to Alabama and Georgia during the fall and winter months. The price paid for apples varies from 20 cents to 50 cents per bushel, and they are sold at Huntsville and other points at from $1 to $1.50 per bushel. Sometimes as many as twenty-five wagons loaded with apples may be seen in one train. These wagons usually take out from 25 to 40 bushels each, making the trip to Huntsville and back in seven days. The wagoners take along provisions and provender enough to last the trip. By this means from $2.50 to $4.00 per day are realized for the use of wagons and teams, and as this trade is carried on when the teams are not required on the farms, it becomes a source of employment and some profit.

Before the war fruit was sometimes sold to parties in Cincinnati, the fruit men paying 25 cents per bushel on the trees. At this time there was very little brandy made, and there were comparatively few orchards.

Since then the fruit farmers in the mountainous portion of the county rank all others, and some very handsome fortunes have been made. Nor is this interest abating in the least. It is extending from the mountain edges to the barrens or flat lands, the returns from the orchards being almost always satisfactory.

The profits on the apples when sold as fruit is very considerable, but the demand is by no means so regular as for brandy, and oftentimes a fruit-raiser cannot afford to wait. About one-third of the fruit is imperfect and not suited to market.

Management of Orchards.—The usual treatment of orchards is to plow the trees twice each year, once in the spring, and again late in the summer. Some field crop is usually planted in the orchard. Corn is generally planted, except when the soil is so thin that it will not pay to cultivate the whole land. In that case the land is plowed for a distance of five feet on each side of the trees. As the orchard approaches maturity the whole of the inter-space is plowed out.

The Limber Twig is rarely ever pruned. The other varieties are trimmed a little for two years. The water-sprouts are rubbed off as they are put out in spring.

The Limber Twig requires no more trimming than a forest tree. The limbs come very near the ground; the twigs are very limber and reach downward. The owners of large orchards seldom ever dry any fruit. This business is confined to small farmers, and tenants living with large fruit growers. It forms a large item in the exports of the county. During the year of 1876 a very large quantity of apples was dried.

Mode of Drying Apples.—They are dried on scaffolds, readily constructed, in the sun, out of rough plank. In good weather the apples will be dry enough in two days to take in and spread in an airy place, under shelter, where, after remaining a short time, they are ready for the sack. The fruit is pared and cut by a machine made for the purpose, consisting of a circular blade. The core is also taken out by the some machine. With this machine one good hand can peel and spread out enough to make one bushel of dried apples per day. About three bushels of green fruit will make one bushel of dry. The Limber Twig makes the best dried fruit, being easy to cook. The average price of dried apples is about $2\frac{1}{2}$ cents per pound, 24 pounds making a bushel. This work is principally done by women and children.

Cider.—Cider is rarely made except for home uses. Made of the Limber Twig in the fall, it can, by pursuing the following course, be kept sweet and good throughout the entire winter:

The cider is made and put in a barrel with the bung left out; as it evaporates add fresh cider, and thus continue from time to time until all fermentation ceases; it is then tightly bunged up. When the cider is drawn it sparkles like champagne, is equally as clear, and resembles the amber hue of wine.

Peaches—are very uncertain. There are a few localities about the mountains where the peach trees bear with some certainty, but as a crop, it has been abandoned as unprofitable and untrustworthy.

The late frosts in the spring are almost sure to destroy the crop, but occasionally it escapes, and the fruit is then very luscious and of the finest quality.

Peaches are generally seedlings, but few budded trees being planted.

Pears—are a very sure crop. The trees are subject to no disease ; they succeed finely everywhere. The principal pears are the Bartlett and Bell. The Seckie is grown to some extent ; all qualities grow well and bear profusely. There are, however, comparatively speaking, few trees planted. Were proper attention given them they would prove quite as profitable as the apple, the trees being very healthy and long-lived. No such thing as "pear blight" has ever yet been seen in the county. Cultivation is unnecessary. I saw trees loaded down with fruit, the soil around which had never been disturbed since they were planted.

Cherries—are also very certain. The May Cherry is not such a sure bearer as the Murillo, the latter never fails ; there is very little attention paid to cherries.

Plums.—The Chickasaw plums bear well and are a great certainty. The Wild Goose plum is more uncertain.

Grapes.—The Delaware ripens to great perfection. The Concord and Ives Seedling do well. The Catawba and Isabella, when protected by the eaves of a house is a pretty sure crop, but they rot and fall off in open grounds. The Scuppernong, contrary to its usual habit, bears well on the clayey soil. A wild grape called the Fox grape, very much like the Scuppernong in appearance, but a little under size, grows wild in profusion. The bunches are small, only two or three berries on a bunch. This grape is said to make a most excellent wine, indeed much better than any of the varieties previously mentioned.

Berries.—Blackberries grow everywhere, rarely ever failing ; a great many are canned.

Raspberries are cultivated in gardens, and are certain bearers.

Gooseberries never fail. Currants, as bearers, are a failure, consequently are not planted.

All varieties of strawberries do well. Hovey's Seedling, Peabody and Wilson's Albany are the principal varieties grown.

Huckleberries grow wild in the barrens and on the mountains. Those grown on the mountains are much larger and sweeter. They grow in great abundance, and are gathered by women and children, and sold to merchants for about 10 to 12½ cents per gallon. Some of them are dried. An active woman can gather a bushel a day, and it forms the principal article of traffic for the women of the mountainous section of the county.

Chestnuts are also gathered by this same class and brought to the market. They generally bring $1 per bushel.

DAIRY INTERESTS.

This is attracting some degree of interest and attention, and where well conducted is proving quite profitable.

Mr. Samuel Mack Ramsey, on Hickory creek, has made an experiment in this direction that is productive of good results, and is likely to be the harbinger of an important industry. In June, 1875, he commenced shipping butter to Nashville for which he got 40 cents per pound in the market. He made then about 40 pounds per week. During the winter he failed to procure a market, so he packed his butter all through the winter, and in the spring of 1876, this butter was shipped to Chattanooga and sold for 30 cents per pound. Fifty pounds per week were furnished to the hotel in McMinnville throughout the summer. During this same summer (1876) the cows and calves were turned together, only taking from them milk enough to supply the engagement with the hotel in McMinnville. Afterwards a proposition came from a firm in Chattanooga to give 30 cents per pound at the dairy for one year. One hundred pounds per week were shipped to this firm, and the demand constantly increased. Two hundred pounds per week could be disposed of to the same firm, and 35 cents per pound the year round, is now offered for another year.

This dairy consists of 30 cows. The best milkers will give from five to six gallons of milk per day, but these are not the best butter cows. Those from whose milk is made the largest quantity of butter, give only two and a half to three gallons per day. Forty-two gallons of milk give 15 pounds of butter. Only the cream is churned. The milk is set in deep vessels stands 36 hours in a spring the temperature of which is 56 degrees. The churning is done at a temperature of 60 degrees. A Blanchard churn is used. It takes about three-quarters of an hour to get the butter. Eight to twelve gallons of cream are put in the churn, and usually from 14 to 18 pounds of butter taken out. The butter-milk is then drawn off, and the butter is washed through three waters of strong brine in the churn. If washed in fresh water the grain and color are both injured. After the washing the butter is taken out and weighed, and one ounce of Ashton dairy salt to the pound of butter is added. It is then put back into the churn and the butter thoroughly worked, until the salt is well incorporated with it. It is then taken up and put in a porcelain butter package, and set in a cold spring, where it can be kept any length of time. It should be re-washed after standing in the spring twenty-four hours. This washing is done by a lever, until all the butter-milk is thoroughly worked out. It is then packed ready for shipping. Mr. Ramsey says he could dispose of 1,000 pounds per week. Good butter is in great demand; bad butter is in abundance, and there is no demand for it. Common butter brings from 10 to 12 cents per pound, while Mr. R.'s butter brings in the Chattanooga market 40 cents per pound and the demand is not half met. Butter-milk is fed to the pigs. Mr. R. thinks it is worth 2 cents per gallon for that purpose.

The calf is taken from the mother at once, but is fed with the mother's milk until three weeks old. A table-spoonful of corn meal is put with a quart of boiling water. The temperature of the water is reduced with sweet skimmed milk, and fed to the calf. The quantity of the meal is gradually increased, until a pint per day is given. This is generally when the calf is two months old—after this time it is fed on sweet and sour milk *ad libitum*.

The cows for 1876, made on an average one hundred and seventy-six pounds of butter each. With better stock, two hundred and fifty pounds to the cow could have been made. Mr. Ramsey milks his cows twice a day, at 6 o'clock in the morning and 4 o'clock in the evening. His cows are treated with great gentleness; stables are cleaned out every day, and the milk vessels are scalded every time they are emptied. His butter is of a golden color both winter and summer. A gallon of meal is fed to each cow twice a day in the winter, and as much herds grass and clover as she can eat. Corn fodder is preferred to any other food. The cows are the best natives, with a few grade Shorthorns. These are better for butter-making than the common cows. In his locality Mr. R. thinks it best to breed his cows to a Shorthorn bull, in order to make beef of the calves. One hand can milk and attend to twenty cows in winter. Mr. Ramsey gives his personal attention to the business, which is doubtless the principal cause of his success.

Grasses for Butter.—Sedge grass, says Mr. Ramsey, through the month of June, will make more butter and better butter than any other grass ; it becomes tough after this, and is almost valueless. This grass, if cut before the straw grows hard makes a very good hay. Horses and cattle are fond of it when cut in bloom. Herds grass is the best grass for a constancy. It lasts longer, and can be grown all the year.

Crow-foot comes up about the middle of May. It grows on wet land, and is very valuable in June and July. It crowds out everything in winter, and is very tenacious of life. Cows and horses are very fond of it. It is very brittle and succulent, and spreads with great rapidity. It gradually exterminates clover and herd's grass. It is the only compeer to broom sedge in its hardiness.

The red-bird clover or Japan clover (Lespedeza striata) made its appearance in 1870, and is covering the whole country. It supplies a large amount of grazing from the first of August until frost. It is short but very hardy. It destroys the broom sedge. Sheep eat it with avidity, but the first frost kills it ; it grows well on red clay soil. Clover succeeds admirably, and is relied on for pasturage from the middle of May until the first of August. Clover is not valued as a butter grass, but as a hay it is held in high esteem. Guinea grass is being introduced as a provender grass.

it can be cut four times during one season; it grows to the height of six feet, making a very rank foliage.

Coal.—Myers' cove, on Panther's creek of Collins river, lies southeast of McMinnville eight miles, and presents a very fine body of farming lands, at an elevation some 110 feet above McMinnville. This cove lies on the edge of the Cumberland mountains, and forms an indentation between two spurs. The limestone is found on the mountain slope 700 feet above the bed of Collins river. Ascending the mountain from this cove, we come to the Keystone Coal Company's bank, which is 750 feet above the level of Collins river. The coal is very much broken and disturbed, and is covered by a heavy bed of blue shale evidently belonging to the lowest coal seam of the Cumberland table-land. The conglomerate has been eroded by time, and no traces of it are seen in this part of the mountain. In the entry made by this company, at the distance of fifty feet from the mouth the coal seam is parted by a mass of shale, one portion rising up at a very sharp angle. The coal of the lower half of the seam is commingled with masses of comminuted shale, and shows much contortion of laminæ. This same seam of coal crops out without disturbance, south of this bank a few hundred yards.

One mile and a half southeast of the Keystone bench, is Barnes' bank, the same seam which crops out at the Keystone bank. The opening of the seam at this place faces a great "gulf," which runs south-east. The coal here pertains to the lower coal measure. The seam is 4 feet thick, and is 95 feet above the limestone, and 90 feet from the top of the mountain. The Fortress or Cliff rock which lies above the next seam, is about 50 feet thick at this place, and the distance between the seams is 35 feet. This interval is composed of shales and thin sandstones. The conglomerate rock is nowhere seen, and the sub-conglomerate seam of coal is also absent. Usually there are three seams of coal below the conglomerate :

1. The lowest shale seam, from 50 to 100 feet above the limestone.

2. The cliff seam, from 50 to 80 feet above the shale seam.

3. The sub-conglomerate seam, lying immediately below the conglomerate rocks. This last mentioned seam, with the conglomerate rock, has been swept away from this locality by erosion.

Twenty miles south-east of McMinnville, on Dry Creek, of Hill's creek, a tributary of Collins river, is a point in Sequatchie county called the Central Coal Fields. Dry creek has a gentle ascent for about nine miles, rising in that distance about 700 feet. The bordering bluffs are composed of sandstone, with wavy and ebb-and-flow structure of strata. Under one of these bluffs, at Hill's saw mill, is an outcrop of block coal, which I am disposed to refer to the cliff seam, inasmuch as it is capped by about 50 feet of sandstone.

8

Six hundred yards east of this mill, on the edge of a little stream, Hill's bank has been opened. The seam here is 5½ feet thick, the coal specimens are fragile but very pure, resembling in every particular the Sewanee coal. The seam has a heavy thickness of bluish shale above, fire-clay and sand-stone below. The seam is 45 feet above that at the mill. I am unable from the data which I have to determine the proper place of this seam. There are two seams below. There is no conglomerate rock, though the hills rise up from 50 to 60 feet above the Hill bank. It certainly cannot be the conglomerate seam, for there is no sign of this rock; besides it is too thick, and the constitution of the coal is entirely unlike. I am disposed to think it belongs to the upper coal measures for two reasons:

1. The coal resembles that of the upper measures.

2. The seam is very regular and continuous, there being no lenticular bulgings or irregularities in the thickness. Several openings have been made in this coal-field, all of which show the same regularity. There must be 10,000 acres here in one body, having a coal seam five feet thick. It is one of the most promising coal fields with which I am acquainted. A railroad could be constructed up the mountain, in Dry creek gorge, with a grade nowhere exceeding 80 feet per mile.

Iron Ore.—Some good specimens of iron ore were shown me, but I could find no extensive deposits of this metal. Some, however, are reported to exist.

Lithographic Stone.—A species of rock much resembling the Bavarian lithographic stone, is found in the county. I had some of it tested, but it proved of an inferior quality. Good stone is thought to exist in the county, and it is believed that a portion of the stone used for engraving Confederate notes and bonds was brought from Warren county.

Hydraulic Cement.—This was manufactured for many years in the county. The quality was considered fair.

Oil Borings.—An earnest effort was made to obtain petroleum in this county, but without success. Mr. Geo. Satterfield, who superintended the borings of the wells, has kindly furnished me with the following letter, which gives an insight into the thickness of the various strata through which the well passed, and their lithological character:

McMINNVILLE, TENN., August 4, 1877.

J. B. Killebrew, Commissioner, Nashville, Tenn:

DEAR SIR—Your favor of the 2nd inst. received, asking record of well-boring in Warren county, Tenn., which you will find below:

Mud Creek We'l.—Seven miles north-east of McMinnville, on Sparta stage road, put down by in 1872 by Tennessee Oil Company ; depth 705 feet.

Sand, gravel and clay.. 20
Hydraulic shale, soft.. 20
Protean or siliceous limestone, very hard...200
Black bituminous shale, soft.. 35
Nashville group, cuts easy...200
Trenton or Lebanon, more compact..230

Whole depth..705

The first 20 feet gave us considerable trouble, on account of the marshy nature of the creek bottom. We had to send to Pittsburg for driving pipe. After putting it in we got along better with the work. Our first rock was a hydraulic shale, cut soft, and is a bluish gray in color. We found the Protean bed formation here to be very different in appearance and character from what it is at the Spring creek wells in Overton county, Tenn. Here it is nearly white and very hard and tough; in Overton it is almost pure silica (blue flint), cuts very hard, but is easy to ream out. The lower 100 feet of this rock I found here to be full of the odor of petroleum, and some light shows of oil. The black shale gives oil shows here the same as in Overton, and is alike throughout the extent of the deposit in Tennessee.

The Nashville rocks lying immediately under the black shale for a depth of 100 feet, is composed almost wholly of marine fossils (molusks) and is soft and easy to drill. It is in this formation that we find the salt water in Warren and White counties. We struck the salt water in our well No. 2, at Rocky river, 11 miles northeast of McMinnville, at 429 feet, and I am told the salt water in the Priest well, and also the Smartt well, in this county, was struck at about the same depth. Large quantities of salt were made at both those wells since the war, but both wells have long since been abandoned on account of the low price of salt.

In the Mud creek well we struck white sulphur water at 489 feet from the surface, that has continued to flow out over the top of the well ever since, but we had no oil shows below the black shale. In well No. 2, on Rock river, we had oil shows in Protean limestone at 125 feet from the surface ; also at 156, 170 and 193 feet from the surface, with the usual show in black shale, and pretty good show in Nashville limestone at 371 feet. This well was put down 525 feet in all. No oil show in Trenton limestone in either well. Character of rock same in both wells.

The Protean limestone formation in Warren county, has but a mere trace of lime in it, and is the most difficult rock to drill I have ever met with ; the borings are white as chalk.

Many years since, some parties sunk a well for salt water on Rocky river

11 miles northeast of McMinnville, and two miles south of where we put down well No. 2, and at about 340 feet in the Nashville rocks, struck oil in large quantities, that flowed out and went off on the river, and was set on fire while flowing. Many old men are living now in this county who remember it well, among whom are John M. Drake, our present Sheriff, and Isaiah Hills.

I am convinced, from my own work and the work of others, that oil, in paying quantities, exists in many localities along the western base of Cumberland Mountain, throughout Tennessee.

Yours, truly,

GEO. SATTERFIELD.